SOLARIZATION

SOLARIZATION

Sandy Walker
Clarence Rainwater

AMPHOTO
American Photographic Book Publishing Co., Inc.
Garden City, New York

Published in Garden City, New York by American Photographic Book Publishing Co., Inc.

Library of Congress Catalog Card No. 73-91379
ISBN 0-8174-0561-5

Manufactured in the United States of America

Preface

This is intended to be a practical book for photographers. The first part explains in detail how to solarize prints, either positive or negative. The second part describes the paper-negative technique and explains how to get special effects, line images, and tone reversals. The third part describes the results of controlled experiments that demonstrate the reproducibility of the solarization process and show the effect of varying the different parameters. The fourth part of the book gives some of the technical background for a deeper understanding of solarization phenomena. And finally, a gallery of prints is offered for enjoyment and to illustrate the range of possibilities inherent in the process.

The original incentive for this book was an accidental discovery made by one of its authors, Sandy Walker, in the summer of 1969. She had left a tray of used developer on the darkroom workbench, and decided to use it for a few prints even though it had turned dark brown. One of the prints was not developing satisfactorily, so she removed it from the developer and dropped it in the trashcan. Later, after turning on the light, she happened to notice that the discarded print was undergoing a very interesting transformation. She quickly snatched it from the trashcan and placed it in the fixing bath. It was somewhat streaked, but otherwise an excellent example of the Sabattier effect, commonly known as solarization.

She then deliberately tried to reproduce the phenomenon, and had immediate success. Upon seeing these results, the second author, Clarence Rainwater, tried to imitate them. Being a careful worker and somewhat compulsive, he started with fresh solutions, not wishing to stain his prints with that old brown developer. The results were disappointing. The prints did solarize, but they also turned a dark, muddy gray, almost black, and showed none of the clear white line and silvery accents that characterized the first examples. Further investigation convinced us that good results required the old brown developer. We began a series of experiments to learn as much as we could about the requirements and limits of the Sabattier effect, and began to use the technique in the production of prints for competitions and exhibits. One of the first things we learned was that we had better call the technique "solarization," correctly or not, if we were to discuss it with our fellow photographers. And "solarization" was also easier to say, so that's the term we used.

The more we investigated and the more we used the process, the more fascinating it became. There were a lot of surprises, and each one brought more possibilities for pictures. Our photographs attracted attention, at least from other photographers, and we became used to the question, "How did you do that?"

Now, two years and hundreds of darkroom hours later, we have here attempted to answer that question and to describe some of the things we have learned about solarization and its uses in creative photography.

Contents

PREFACE 5

INTRODUCTION 9

1. *SOLARIZING THE PRINT* 13
Selection of Materials and Arrangement of the Darkroom 13 • The First Print 17

2. *THE PAPER NEGATIVE* 23
The Paper Negative Process 23 • Paper Negatives and Solarization 30

3. *SPECIAL EFFECTS* 45
Negative Sandwiches 45 • Double Printing 49 • Repeated Images 55 • Line Drawings 58 • Block Prints 63 • Etchings and Texture Screens 65 • Bromoil 73 • Turning Day into Night 76 • Turning Night into Day 78 • Bas Relief 80 • Tone Separation 83 • Gentle Solarization 87 • Bleaching 90 • Adding Color to the Image 94

4. *SOLARIZING COLOR PRINTS* 97

5. *THE SCIENCE BEHIND THE ART* 103

A GALLERY OF SOLARIZED PRINTS 129

BIBLIOGRAPHY 157

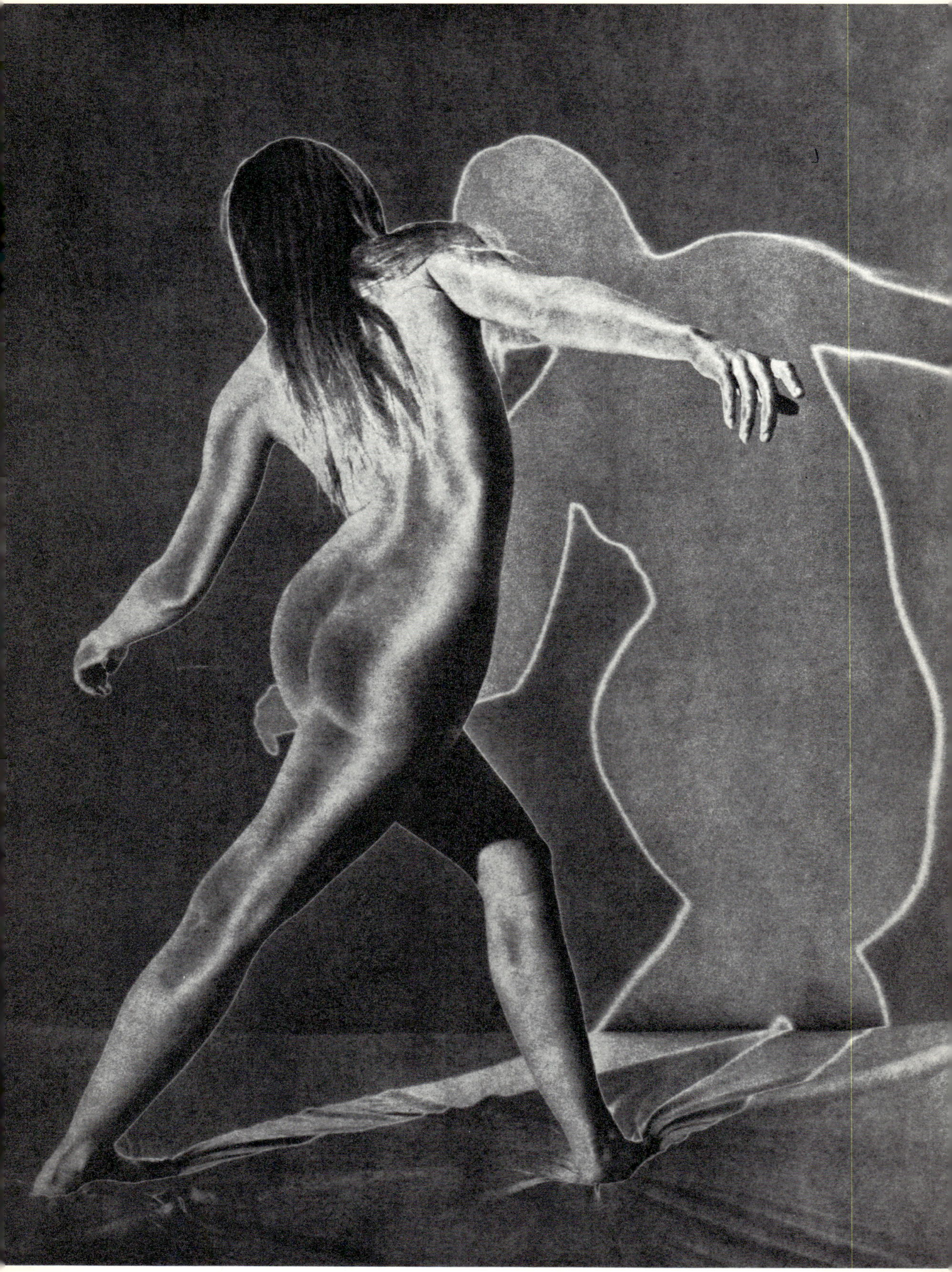

Introduction

The photographic process contains a number of mysteries that are not yet completely explained and that promise challenging work for another generation of photographic scientists. Many of these are simply called "effects" and are most often named for their discoverers. The Sabattier effect is probably the most widely known of these, although the majority of practical photographers call it "solarization." This complex phenomenon is a partial or complete reversal of the photographic tones when the sensitive material is reexposed to light during development. It was discovered in 1862 by the French scientist, Armand Sabattier, but it first achieved notice as an artistic device in the work of photographer-artist Man Ray in 1929.

In technical literature, the word "solarization" is applied to the case of extreme overexposure, when the density of the developed emulsion actually decreases with increased exposure. While it is possible to achieve a tone reversal by such overexposure, we wish to make it clear that none of the pictures displayed in this book have any connection with this true solarization. Our work is entirely concerned with the bogus variety, pseudo-solarization, the Sabattier effect. Like other practicing photographers, however, we will, for simplicity, call it solarization.

The process is briefly described in a number of photographic how-to-do-it books, and is occasionally reviewed in the popular camera magazines. In most of these accounts, however, the emphasis is on solarizing negatives on film, and the reader is warned that the results are likely to be erratic and unpredictable. If such were really the case, we would not bother to write this book. The fact is that you can solarize prints on ordinary enlarging paper with results as good as you can get on film, and the whole process can be carried out under a yellow safelight, watched and controlled just like any other printing process. What is more, these solarized prints may be used as paper negatives and printed again to produce positives or negatives, as you wish, or even be solarized a second or third time. Furthermore, because you are working with a large piece of paper instead of a small piece of film, you can draw or paint on the picture to modify it in any way that you please. The final print may then be a real work of art, and not just a record of the scene in front of your camera at the moment the shutter was tripped. In short, this combination of solarization with paper negatives endows photography with a creative potential unexcelled by any other medium. Your pictures may resemble drawings, etchings, charcoals, bromoils, or woodcuts. On the one hand, the images you create may illustrate dreams, fantasies, or nightmares—visions of the mind's eye rather than the camera's—and on the other be stamped with that convincing truthfulness that is the special province of photography.

Some photographers, and a number of critics as well, believe that there is a certain integrity in a "straight" photograph that is violated by any form of control process. They deny that any artistic improvement can be achieved by "non-photographic" manipulation of the image, claiming that it always produces a contrived and artificial appearance. We feel that this is a reactionary attitude, adopted as a defense against the charge that the photograph is too literal to be "art." Instead of refuting the accusation, they have accepted the charge as true, but claim the literal representation as a virtue essential to the medium. We hold that the accusation is false—the photograph is not necessarily literal. Its acclaimed veracity is the deliberate result of successful effort on the part of photographic scientists and lens designers to produce a recording instrument for commercial and technical purposes. In creative art, the use of a camera is as incidental as the use of paint or a palette knife. If all artists were to vanish tomorrow, the photographic and paint industries would continue with hardly a nod. In our view, the photographic process is simply a way of fixing an image, as abstract or as realistic as the artist chooses, and there are no holds barred in the production of that image. In the end, the picture must be its own justification, and like all works of art, be finally judged by the individual viewer according to the response it evokes in him. To limit oneself to the images produced by the camera is to accept an unnecessary discipline, like a painter who only uses three colors. Success may provide a puritan satisfaction, but it does not vitiate the work of the artist who uses a full palette.

In the popular literature of photography, the subject is surrounded by an amazing number of warnings and admonitions. The list of "do's" and "don'ts" is almost endless, and there are precise prescriptions for doing everything from pressing the shutter release to mounting the print. While these pronouncements are aimed primarily at the novice, they are often so thoroughly ingested that they continue to restrict his work after he has become more experienced. A photographer so conditioned may make the same kind of pictures year after year with little change and not much improvement. Although his prints are sharp and clear, he fails to realize the impressive dimensions of the medium and entirely misses the exciting adventure of the unexpected.

The instructions in the how-to-do-it manuals (including this one) are only guides along known paths to known results. The pathway to creative photography, however, is uncharted, and leads through the wilds of experimentation. The primary requirements are persistence and awareness. When you experiment, pay close attention to exactly what happens, make notes so you will remember what happens, and you cannot fail to learn more than you ever thought there was to know.

Many of the photographs in this book contain nude figures. We feel that the naked body is not only a sensual and moving image, but also a primitive and powerful elemental symbol. Clothes, like automobiles and other artifacts, are stamped with a date, framing the picture within an era and a culture, and should not be used unless such labeling is desired. It is in relation to nudity that the camera has been most unjustly criticized for its literal representation. Critics have frequently experienced such panic before a photograph of a naked human being that they have been unable to assess its artistic merits as a picture. Many a young photographer has been cowed into inaction by dire warnings that in photographing a "nude" he might inadvertently perpetrate a work of "bad taste." The nude is often said to be the most difficult of all photographic subjects. We say that that is nonsense! It is no more difficult to make a good photograph of a naked body than to make a good photograph of anything else, a face, a car, or a cat, for example. And if you do make a bad photograph, it is only a bad photograph, no matter what the subject, and not a "transgression of the bounds of good taste," whatever that may mean. If your artistic sensibilities are low and untrained, you won't make very good pictures of anything, nude or otherwise, and furthermore, you will never learn to, unless you try again and again. One last point: in spite of the many condescending instructions and admonitions about the proper treatment of figure models, there is really nothing to fear. A nude model is just like any other model, and responds just as well to being treated like a human being. In this book we have tried to illustrate the variety of solarizing phenomena with many different subjects, but our special delight has been the interpretation of the mood and form of the human figure.

Our pictures have been made from negatives taken over a period of several years, using at least fifteen different cameras and six different film sizes. We keep records of the subject, date, camera, film, and development data, and also note the general character of the lighting, such as electronic flash, flood, or daylight. We do not record shutter speed and aperture settings unless they are unusual. Our negatives are of widely varying densities and contrasts, the usual criteria of quality not being applicable for the kind of work we like to do.

1

Solarizing the Print

SELECTION OF MATERIALS AND ARRANGEMENT OF THE DARKROOM

Different photographic emulsions, whether on paper or on film, respond differently to reversal by solarization, although nearly all of them will show some effect. Most published descriptions refer to negative emulsions on film. The solarizing exposure on panchromatic film can be made by turning on a yellow safelight for about ten seconds while developing the film in darkness. The negatives obtained are invariably of low contrast, very dense, and require long exposure times when printing. The results are often referred to as unpredictable, and success is the result of trial and error or, more often, a lucky accident. Part of the trouble, we suspect, is that you can't see what you are doing in the dark. Solarizing the print instead of the negative is not usually discussed—for who wants a low contrast, very dense print? But there are very great advantages in working with enlarging paper instead of negative film: You can leave the yellow safelight on and see clearly what is happening. The solarizing exposure is made with an ordinary light bulb, and because the image is a positive one, you can see your picture directly and control the results to your liking. The problems of contrast and density can be overcome by the use of a well-seasoned developer. It is the purpose of this chapter to explain how this is done.

The solarized print is ideally suited to the paper-negative process as described in the next chapter. It is very easy to make copies and variations of the print with complete control over tones and contrast. Because it is simpler, and because it affords such complete control, we will consider only the solarization of prints on paper. We should mention, however, the possibility of using positive copy film, which can be handled under a safelight. Such film can be solarized by the procedures described here, but it is not as easy to observe the process because of the milky coating that must be dissolved in the fixer before the film becomes clear. The film may be treated in the same solutions and handled just like paper, but it is more subject to damage by scratching and does not lend itself to the extensive handwork possible on a large print. Furthermore, the solarized film itself is not suitable for exhibition, but must finally be printed on paper. This is a serious limitation, because one of the most impressive features of the solarized image, either positive or negative, is the delicate white outline (Mackie line) emphasizing the design. In the print from a solarized negative, this outline is black, and not nearly so attractive. The only advantage to be gained by using film is the ability to enlarge the image to any desired size.

The best paper for solarizing is a high contrast, bromide enlarging paper. We have found Agfa Brovira No. 6 to give very dramatic results, and we use it for about eighty percent of our work. It is the "hardest" paper obtainable, gives very clear whites, deep rich blacks, and solarizes with poster-like clean lines. Papers of low contrast are useful as paper negatives and for certain types of pictures, but they are often too dark for satisfactory display prints. Portrait papers do not respond very well to solarization, having a tendency to simply turn negative without any desired edge effects. So, choose a bromide paper of contrast grade No. 4 or higher, preferably single-weight, either glossy or smooth. Single-weight paper will give better results than double-weight when used as a paper negative because its light transmission is higher and more uniform.

Perhaps the strangest advice we have to give concerns the developer. Use an ordinary paper developer of the D-72 type, such as Kodak's Dektol, GAF's Vividol, or Dupont 53-D. Dilute it with two parts water as you normally would, and then age it until it is old, tired, and brown! By aging, we mean develop a lot of prints in it, perhaps forty or more 8″ x 10″s. It will turn brown, look like weak coffee, and take about twice as long to develop a print as when fresh. The process can be hastened by leaving a sheet or two of printing paper in the tray to turn black. In other words, the developer is ripened by reducing a lot of silver in it, so whenever you have made a printing mistake, just leave it in the developer to help with the seasoning process. This old brown developer is necessary for successful print solarization. Fresh developer works too fast to allow control of the process, and also fails to produce the luminous whites attainable with "old brown." Very dilute fresh developer will work slowly enough, but it will not produce good blacks. Instead, it gives brown tones characteristic of incomplete development.

Do not be afraid that the brown developer will stain your prints. If you use a fresh stop bath and fixer, you will have no trouble with staining. In fact, the aged solution can be used to develop your straight prints as well as the solarized ones. It will act more slowly than fresh developer, but you will not find the quality of the results significantly diminished. It is a good idea to use a black developing tray, because the wet black tray can be used as a gauge to compare with the blackness of your developing print. Following development, a solarized print is treated as any other in a standard acetic acid stop bath and a reasonably fresh fixing solution. After thoroughly washing the prints, place them in a good print-flattening solution for about five minutes before drying. This is very important if the prints are to be used as paper negatives, because any curl or edge crinkle will make them difficult to handle and may result in unsharp or mottled prints.

In addition to the usual pieces of equipment in your darkroom you will need two sheets of plate glass slightly larger than the prints you plan to make. For 11″ x 14″ prints the glass plates would be at least 13″ x 16″, which allows extra room around the edges so that you can hold them tightly together without obstructing the light. Ordinary window glass will not do—it is not flat enough. Use good quality plate glass 1/4″ thick. Tape the two pieces together along one of the long edges so they will open like the covers of a book.

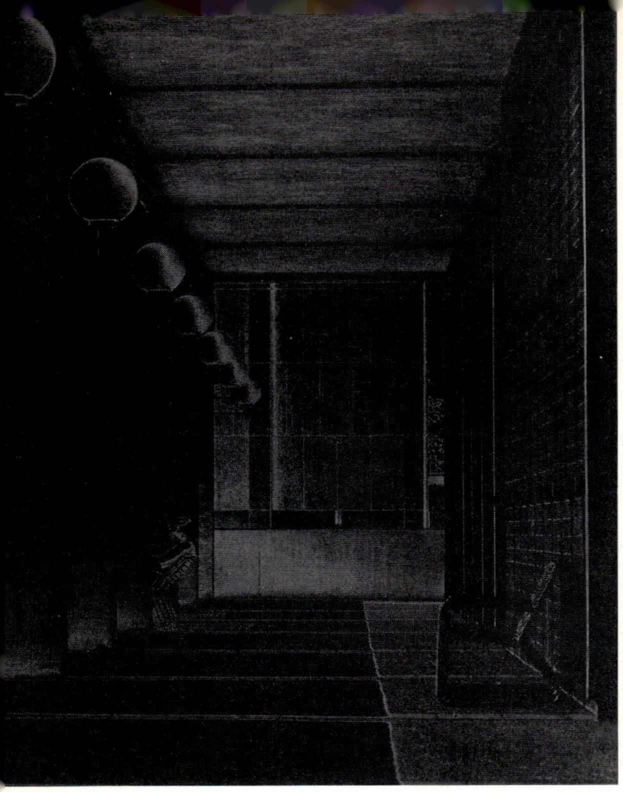

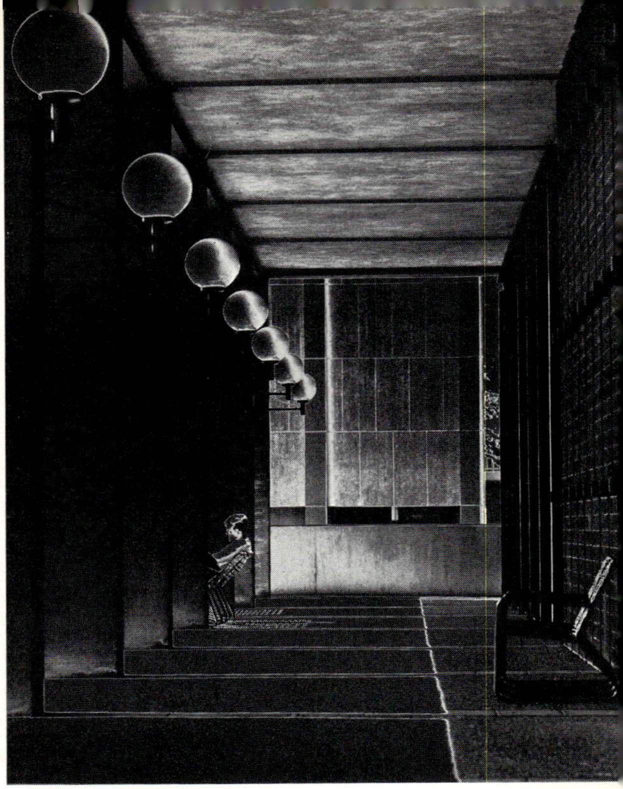

1-1A 1-1B

Figures 1-1A & B. Figure 1A shows a typical result of solarizing a print in fresh developer. The print has been given a normal exposure on Brovira No. 6 paper, developed and solarized in Dektol diluted with two parts water. Figure 1B received the same treatment as 1A, except the developer was well-used and brown. Note the luminous whites, the generally more successful solarization.

Lay these pieces of glass on one end of your workbench, some distance from your developing trays, so they will not be splattered by solutions—next to the enlarger is fine. About three feet directly above them mount a 25-watt light bulb operated through a convenient switch. A foot switch is useful here, or you can connect the light through your enlarger timer. Arrange a shade around the bulb to restrict the light to uniform illumination of the glass plates. This is all the equipment you need to make prints from paper negatives or to make paper negatives from prints.

For solarizing you will need a white light bulb directly above your developing tray. Use a 25-watt bulb at a height of three feet or a 60-watt bulb at a height of five feet. The solarizing exposure will usually be not less than two seconds long nor more than ten seconds long, and is actually not very critical. An automatic timer is not necessary, but a foot switch or a simple pull cord is convenient, since your hands may be wet. Convenience is the main consideration, and should your actual arrangement differ from the one described here, slight changes in your exposures will easily compensate.

THE FIRST PRINT

While solarization will often make a picture more interesting and dramatic, it is not a panacea guaranteed to turn a bad picture into a good one. Keep in mind that the usual result of solarization will be an emphasis on the design of the picture at the expense of representation. Texture and modeling will be sacrificed in favor of outline and tonal contrasts. There will frequently be a fusing of the background and foreground, with a consequent loss of perspective. The final picture may look more like a poster than a photograph, and have the same startling effect of unreality.

For your first experiment in solarization, choose a good sharp negative, having at least average contrast but not too much fine detail. Using No. 6 grade paper, make test strips to determine the exposure time such that the picture develops fully in 2-1/2 minutes in the well-seasoned "old brown" developer. Make a straight print on this high contrast paper, and keep it at hand to compare with the solarized versions.

Expose another sheet of paper exactly the same way. Develop this for about 45 seconds, or until all of the main features of the picture have appeared. Then turn on the overhead white light for two seconds. Rock the developer tray during this exposure to keep any sediment moving, and be careful that no part of your body or the print tongs casts a shadow on the print. Now turn off the white light and watch the print. You will see the whites begin to darken, rapidly overtaking the already developed areas, and finally becoming the darkest parts of the print. A silver blush spreads over the middle tones, and a white line traces the outline of the image. There are few moments in all of photography as exciting as watching this transformation take place. It will remind you of the first time that you watched a developing print appear as if by magic. As the reversal takes place, the image tones will momentarily fuse as they pass through the equilibrium stage of matching density, and then they will exchange roles, the shadows becoming the highlights and the highlights becoming the shadows. You may have the sensation that the picture is turning inside out, the solarized parts seeming to come forward while the original image recedes.

Continue the development for at least 45 seconds after turning off the white light. Resist the temptation to snatch the print out of the developer to save the whites from turning black. If the printing and solarizing exposures have been properly done and if the developer is well aged, the image will not disappear into the mud even if the second development time is extended. A shortened second development will produce a brown-toned image with silver shadows and good detail, but only a partial reversal will occur. During the early stages of the second development the print is very sensitive to uneven distribution of the solution, and the tray should be gently rocked to prevent streaking. If you decide to remove the print from the developer at this stage, be very careful not to touch its surface either with your fingers or print tongs except at an already black area that is fully developed. Remove the print quickly and immerse it immediately in the stop bath, because draining it will invariably produce streaks.

A longer second development will produce a black image with a fine silver outline and silver highlights where the border region between the highlights and shadows used to be. In the case of a back-lighted subject, the result may be practically a silhouette drawn in silver on black.

If the print has been underexposed in the enlarger, the solarized image will be incomplete, and the original highlights will be uniformly black without detail. The reversal will be nearly complete, however, and the image will be mostly negative. A similar result will be obtained if the print is properly exposed in the enlarger but the second exposure is given too early in the development. In both cases, unless the image has developed until it is visible under the safelight, it will be swamped by the second exposure and the whole print will turn black.

If the print has been overexposed in the enlarger, the image will show little reversal, but may develop a white outline and a gradation of tones in the highlights. This will also be the case if the print has been properly exposed in the enlarger but the second exposure is given too late in the development. In general, the second exposure will inhibit the development of the first exposure and the first development will desensitize the emulsion to the second exposure. Many different effects can be achieved by changing the first and second exposure times and the first and second development times. Chapter 5 will deal with these variations and explain how they may be used pictorially.

Depending on the negative and the technique you have chosen, the solarized print may look like a negative or a positive or some strange combination of negative and positive. This predominantly black print with silvery highlights that you have produced may be chosen as your final result and prepared for exhibition. However, this does not end the possibilities. The solarized print itself makes a very fine paper negative, and the next chapter will describe how to use this technique to obtain further derivations of the solarized image.

Figures 1-2A to F. A series of six prints of a water tower illustrating the simple derivations easily obtained from a single negative.

1-2A. A straight print on Luminos paper, grade F-3. The picture was taken with a Pentax on Panatomic-X film in mid-afternoon sunlight.

1-2B. A print on Agfa Brovira No. 6 paper, given normal exposure in the enlarger and solarized midway in the development. Note that the highlights and small shadows are reversed, but the large black shadow on the side of the chimney is not.

1-2C. Turning print 2A over and printing through the back produces this version. All gray tones have been eliminated, leaving solid black lines on a white background, with fine scattered dots to indicate shading.

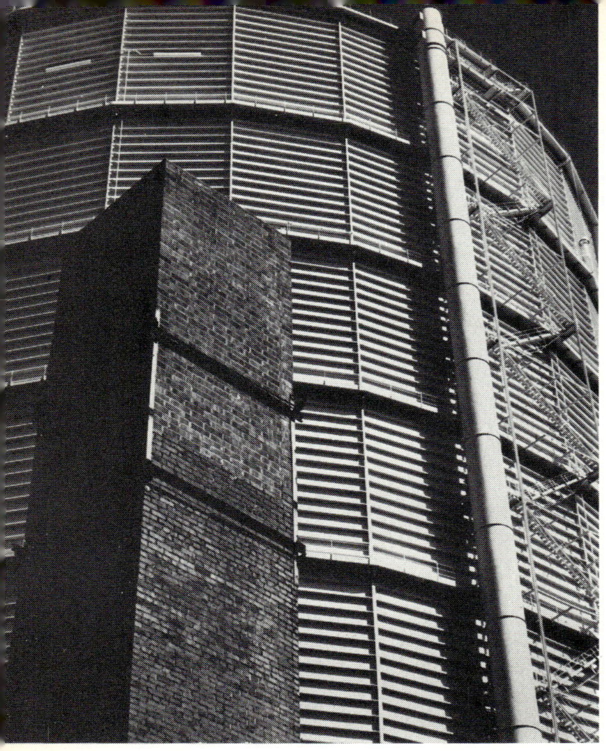

1-2A

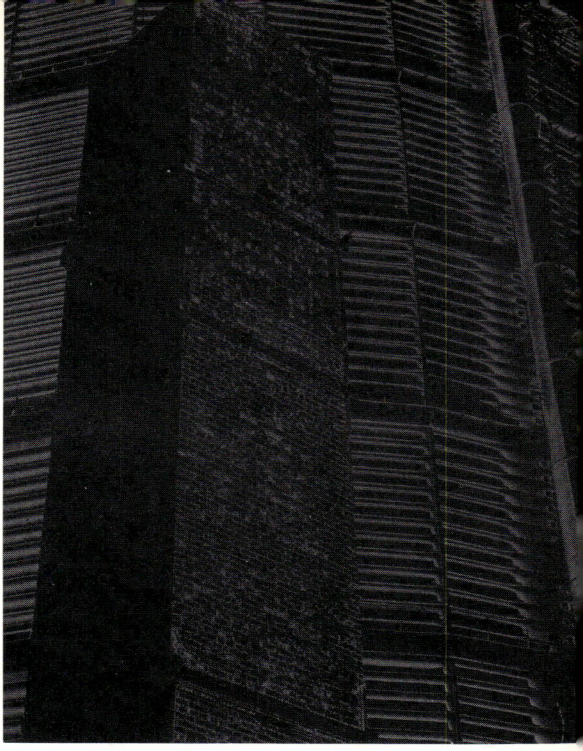

1-2B

1-2D. Printing 2C in turn gives a print which is similar to 2B, but the silver-gray tones have now become pure white. This effect is one type of posterization.

1-2E. This is a double solarization. It is the result of solarizing a print like 2C. Compare it with the single solarization represented by print 2B. There is now a double Mackie line around the details, and the ladder on the side of the tank is no longer recognizable.

1-2F. This is a print from 2E, exposed just long enough to give light gray areas where the highlights were originally. Compare with print 2C. Note that the regular pattern of the bricks is now beginning to disintegrate. Whereas print 2C somewhat resembles a pencil drawing, this is more like a pen and ink drawing.

We call our first solarized print the ''original'' or ''zeroth'' generation. The next print from that will thus be the first generation, and printing the first generation gives the second generation, and so on. In the above series, we ignore the straight print and call print 2B the original. Then prints 2C, 2D, 2E and 2F become the first, second, third, and fourth generations, respectively.

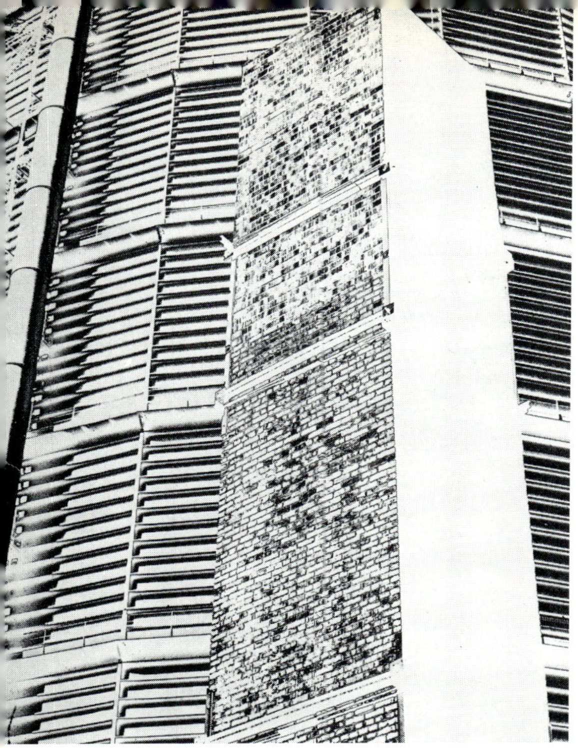

1-2C

1-2E

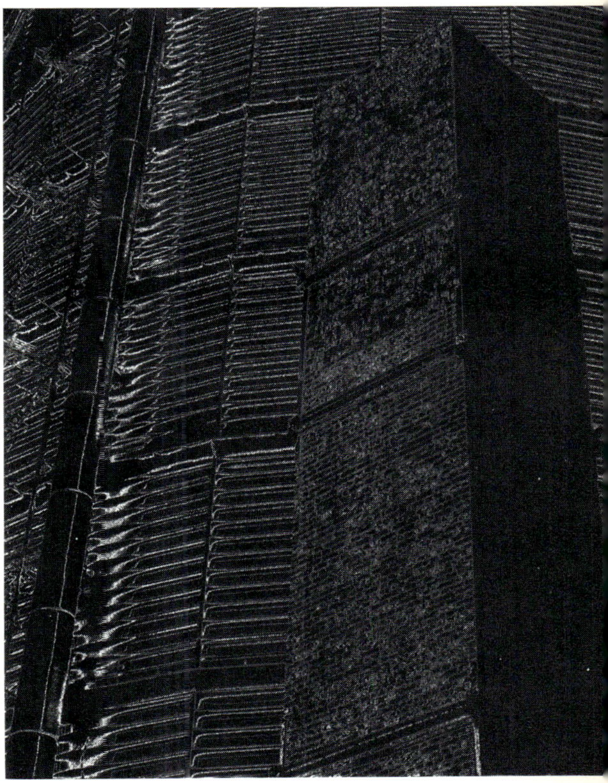

1-2D

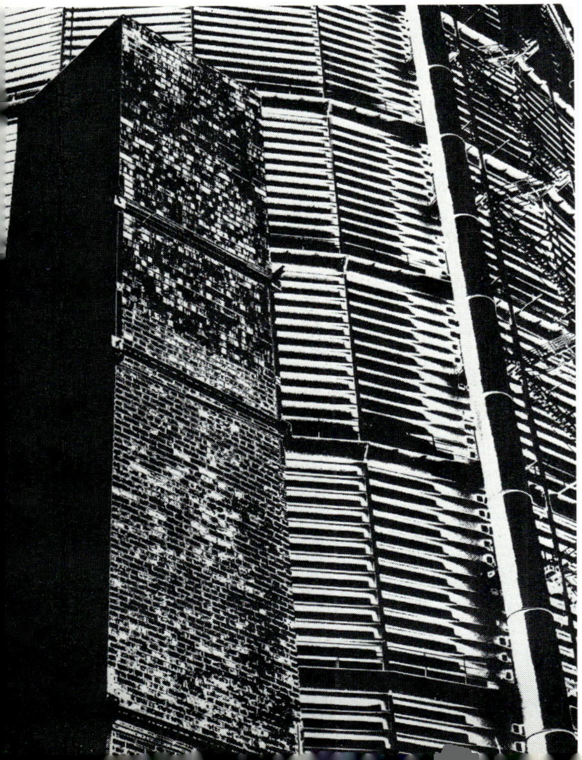

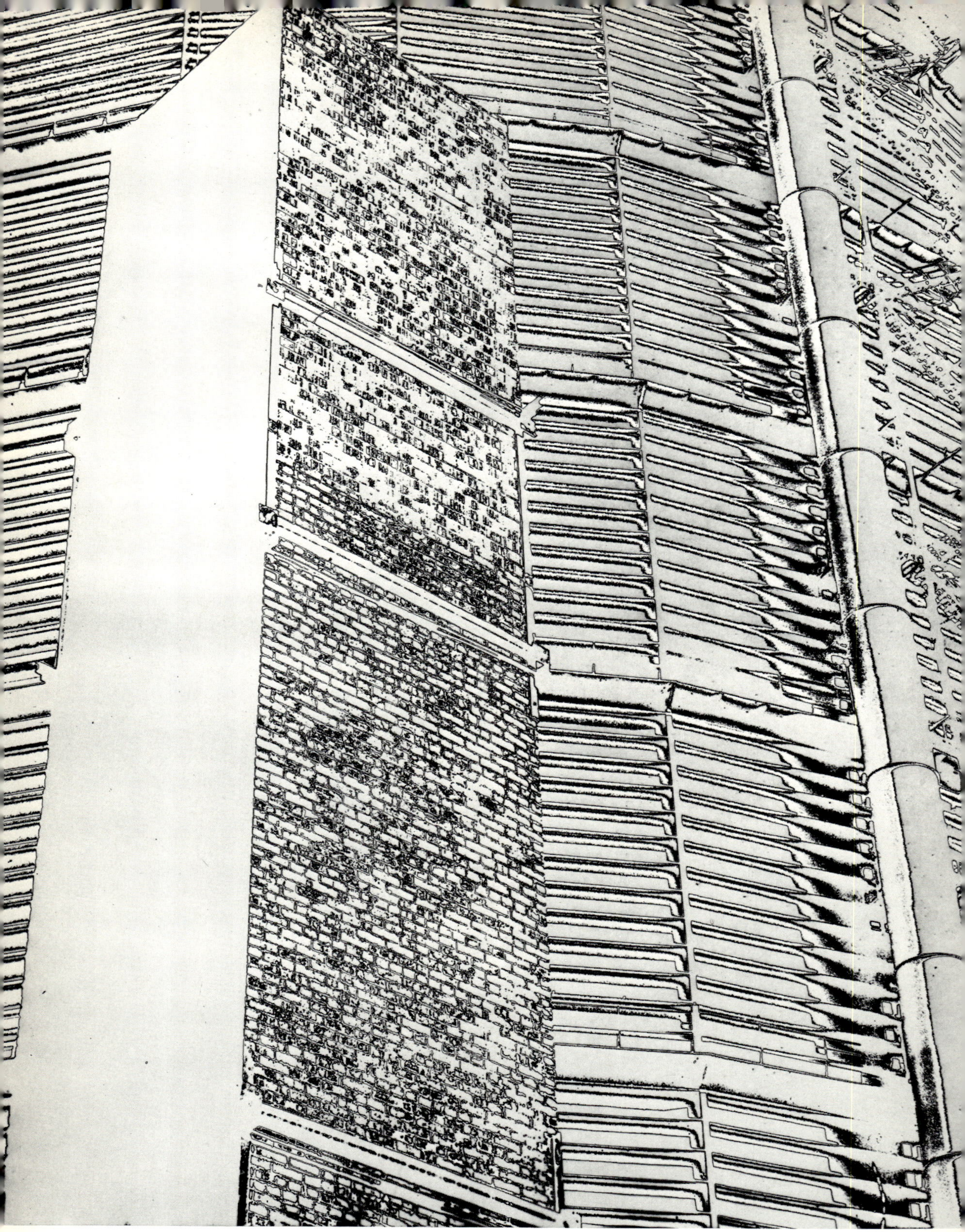

1-2F

2

The Paper Negative

THE PAPER-NEGATIVE PROCESS

The paper-negative process is almost as old as photography itself, having come into being in 1835, four years before the advent of glass-plate negatives. The first paper negatives were made by William Henry Fox Talbot, who exposed silver chloride paper directly in the camera for half an hour. His negatives were only one inch square, but he later made them larger and faster, and finally patented the process under the name of Calotype. This was the first process in which a negative was used to print a positive image, and it had the advantage over the direct-positive daguerreotype in that it could make an unlimited number of copies. It should be noted that the coated paper was ordinary writing paper, never intended for photographic purposes, and was consequently not as uniform in texture as our modern printing papers. After enjoying considerable popularity during the 1850's the direct paper negative and the daguerreotype were finally abandoned as more and more photographers turned to the faster wet collodion plates.

Although no longer exposed directly in the camera, paper negatives have survived as a useful intermediate step in the production of pictorial photographs. During the era of large cameras, retouching and more extensive modification of the image was done on the large film or glass-plate negatives. Since many of the early photographers were trained as artists, their photographs were imitations of romantic paintings, and there was a continuing controversy over the question of whether or not photography was "art." These romantic pictorialists were severely criticized by the purists, who held that the unretouched silver image is the only true aim of legitimate photography. The introduction of smaller and smaller cameras worked in favor of the purists—it became more and more difficult to do handwork on the negative, and many a purist was born of the inability to retouch his negatives.

It is here, in the modern age of 35mm film, that the paper negative regains its place of importance. If you are not a purist, or are ready to quit being one, here is a way to take control of your pictures, and overcome one of the greatest disadvantages of the small camera.

As a technique, the paper-negative process has sometimes enjoyed a wave of popularity in one era only to be forgotten in another. During the 1930's and early '40's, articles describing paper-negative work appeared rather often in photography magazines, but by 1950 it had apparently fallen out of fashion. It is surprising how many modern photographers are not even aware of this very useful and uncomplicated technique.

In principle, the procedure is quite simple. A print is placed face down in contact with a piece of sensitized paper and exposed through the back to white light. The paper is then developed, producing a negative image that, after fixing, washing, and drying, can be printed in the same way to produce another positive. The great advantage of the process is the amount of control and image modification that can be introduced by handwork on the original print and on the paper negative. The original print is called the diapositive, and its quality will greatly influence the tones of the final print. If the final print is to be of full scale, with a good range of tonal values, the diapositive should be of low contrast and dark enough to contain all wanted details. Subsequent printing steps tend to build up contrast, so the diapositive should be printed on a softer grade of paper than a normal print if normal contrast is to be retained. This technique is particularly effective for improving negatives that are too thin or for which the lighting was too flat. It cannot rescue a negative that has no detail in the shadows and has blocked-up highlights.

Ideally, one would choose a printing paper with a smooth or semi-matte surface for both the diapositive and the paper negative, for one could then do handwork on either side of the paper. In our work, however, the diapositive is often made on glossy paper because that is the only surface available with the very high contrast that is best for solarizing. Black drawing ink (the kind used on acetate) can be used on the front of a glossy print for line work or opaquing, but all other handwork must be done on the back. A single-weight paper is better than double-weight, both for transparency and uniformity of texture. It is not necessary that the diapositive and the negative be made on the same kind of paper, and, of course, the final print can be made on any kind of surface you choose.

Many photographers do not consider the use of paper negatives because of the mistaken belief that printing through paper will produce an unsharp picture, or at least a very grainy one due to the paper fibers. There will be no trouble with sharpness if all the exposures are made emulsion-to-emulsion, and the papers are held in good contact during the exposure. With some papers, however, it is true that the texture of the paper fibers will show up in the print as a faint mottling effect. The effect is due to the uneven texture of the paper fibers and the resultant uneven exposure when the light passes through the paper to reach the emulsion. For certain types of pictures this mottling is not at all unpleasant but adds to the mood of the print, giving an effect similar to that of a texture screen. If it is undesirable, however, it can be minimized or eliminated by a simple trick: a very brief flash of light to the back of the print either before or after imprinting the latent image on it. The enlarger, without a negative in the carrier, is used for the flash. This uniform flash of light to the back of the diapositive and to the paper negative will slightly fog the paper in a mottle corresponding to the texture. By reflected light this mottle will be evident, but by transmitted light the variation in the opacity of the emulsion will compensate for the variation in transparency of the paper. The exposure from the rear must be determined by test, but it is not as critical as one might think, and is approximately a constant for any given paper. Fortunately, in the case of creative photography such as we are concerned with here, there is little problem with either lack of sharpness or with fiber grain. The fiber grain can be either suppressed as described or it can be included as an aspect of the overall picture design.

It is very important that the print lie flat after drying, since curl and edge crinkle will make it difficult to use. Curling will prevent the two sheets of paper from lying together in good contact, and crinkle on the edges will show up as dark scallops on the subsequent print due to light leaking around the edge. A thorough soaking in a print-flattening solution just before drying is recommended. A satisfactory mixture is made by mixing one part glycerin with eight parts water. After the print is thoroughly washed, immerse it in this solution for at least five minutes, then lay it face down on a clean ferrotype plate, glass sheet, or other hard surface, and remove the excess water from the back with a squeegee or a paper towel. Commercial solutions, such as Pakosol and Kodak Print Flattener, can be used in much the same way as the glycerin mixture, and their manufacturers claim they give superior results. The print can then be dried in a commercial drier, or in a blotting roll, or just laid face down on a clean blotter. In the latter case, there may still be a slight curl when the print is completely dry, so it is advisable to let the last bit of drying take place under pressure. A flat board weighted with a stack of magazines makes an adequate press.

The dry diapositive is now placed face down on a light box or similar viewing device to be inspected by transmitted light. (See Fig. 2-2 for details of light box.) The image is studied carefully and decisions are made regarding the tones and details of the final picture. This is the point at which the photographer exercises his creative imagination. He decides what parts of the final print are to be darker, what parts are to be lighter, what is to be added, and what is to be deleted. It is very helpful to draw a rough sketch of the picture, with notes regarding the changes to be made. All changes involving an increase in density of the final print should be made on the diapositive, as most of this work is additive, using dark pigments. Subtractive work, as with bleach or an etching knife, is more difficult and seldom necessary. All changes involving a decrease in density of the final print can be made by adding density on the paper negative.

Retouching methods will be described in detail later in

this chapter. For now, however, we are concerned with the general procedures of making and printing paper negatives. This is a useful technique even without the possibility of retouching, simply for the ease with which a negative can be turned into a positive, and vice versa. Here is where that plate-glass printing frame described in Chapter 1 comes into use.

In the darkroom, with only the safelight on, open the glass frame and place a piece of enlarging paper on the bottom glass face up. Place the diapositive face down on top of the enlarging paper and slowly lower the top glass into place, being careful to keep the papers in register. With your fingers near the edges of the glass, press down firmly with both hands, so that the diapositive will be held tightly against the paper underneath. This pressure is very important, because any failure of contact will cause a fuzzy area in the print. This is also the reason that the glass must be thick and flat. A thinner glass would bend under the pressure and not give good contact. Since both hands are now occupied, you will have to use the foot switch to make the exposure with the 25-watt light bulb above the frame. The proper exposure will have to be determined by test, but it will probably be between five and fifteen seconds. To make a test strip, use a large piece of opaque cardboard to shield 3/4 of the paper. Expose for eight seconds. Move the cardboard to expose 1/2 the paper. Expose for four seconds. Move the cardboard to expose 3/4 of the paper, and expose for two seconds. The last exposure with the cardboard completely removed is also for two seconds. This gives a series of exposures of 2, 4, 8, and 16 seconds. Develop for a full two minutes. After briefly fixing and rinsing the test strip, hold it up before a white light and examine it by transmitted light. Remember that this is a test for your paper negative, and some portion of the strip should look like a good negative. There should be no completely opaque or completely white areas, and all wanted details should be clearly defined. After you have chosen the correct exposure and printed and processed your paper negative, turn it over and make a test print from it. Examine the test print carefully, and decide what further changes are needed. Since you have already taken care of the shadows by working on the diapositive, you will probably find that it only needs some of the highlights lifted. This can be done by placing the paper negative face down on the light box, and darkening the desired highlight areas with a pencil or brush. The same tools and techniques for working on the positive can also be used for the negative. These are explained and illustrated later on in this chapter.

In order to save time, you may sometimes want to print a paper negative while it is still wet. This can be done successfully provided the paper negative has been washed to clear it of fixer and care is taken to insure that no drops of water remain between or on top of the two sheets of paper. The best way is to immerse the unexposed sheet in the wash water with the negative and place them face-to-face before removing them from the water. If you take the wet negative out and place it in contact with a dry sheet of unexposed paper, you run the risk of having the emulsions stick together as the dry sheet soaks up water from the wet one. Lift the two sheets out of the water together, drain off the excess water, and place them in position on the bottom glass plate of the printing frame, remembering to have the negative on top. Using paper towels, press the two firmly together and thoroughly wipe the water from the top surface before bringing the top glass down and making the exposure. Wet paper is somewhat more transparent than dry, so the exposure time must be reduced accordingly. The fiber texture will be more evident when a print is made from a wet negative, giving a mottled effect that may or may not be desirable. It will be most noticeable in large gray areas such as the sky. In very high contrast work, where everything is either white or black, there is no trouble with mottling. In most cases, however, it is preferable to wait a little longer and print from a dry negative. In Figs. 2-1A to I we show the type of fiber-grain pattern that results from printing through different dry printing papers.

Figures 2-1A to I. Test for paper grain effect. Paper fiber patterns made by printing through unexposed but fixed, washed, and dried enlarging papers. The neutral grey density was chosen to emphasize the grainy pattern.

2-1A

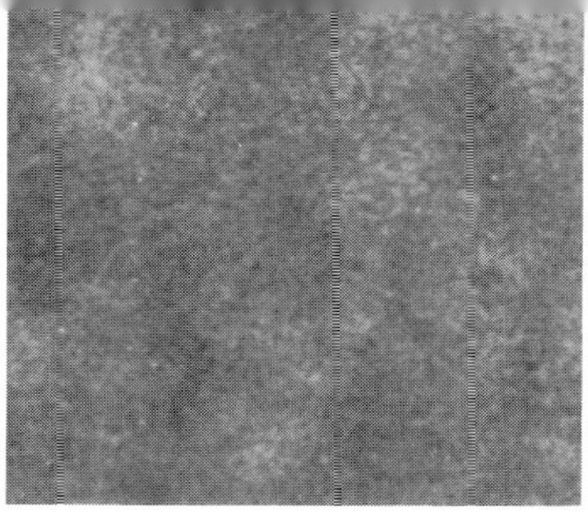

2-1B

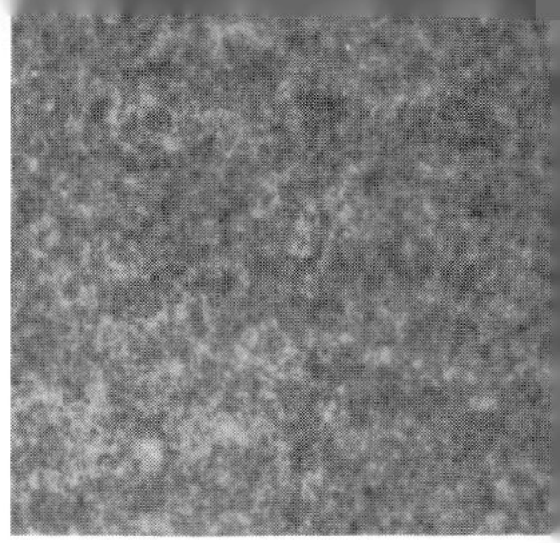

2-1C

2-1D

2-1A. Single weight Medalist F-3.

2-1B. Double weight Polycontrast N.

2-1C. Luminos double weight Charcoal Rough.

2-1D. Luminos Portrait Proof Rough No. 2.

2-1E. Varigam single weight RW.

2-1F. Luminos Bromide Commercial F, single weight, No. 3.

2-1G. Agfa Brovira single weight 1, No. 6.

2-1H. Mural single weight R-2.

2-1I. Agfa Brovira 119, double weight, No. 3.

2-1E

2-1F

2-1G

2-1H

2-1I

The previous description is intended to guide the reader in the general use of the paper-negative process, and has actually been given in more detail than necessary for our present purposes. Once you have experimented a bit with printing on paper through paper, you will see how easy it is to achieve unusual results, and how the results can be controlled to your satisfaction.

Flopping a print over to make a paper negative or vice versa will become commonplace, and you will find it quite natural to take up a pencil now and then to modify some detail of the image. Since the prints you will obtain by solarization are neither quite negative nor quite positive, you will have the option of choosing the first solarized print or any of its printed offspring as your final result. In most cases, the solarized print will be considered to be the diapositive, since it meets the requirements of high density and low contrast. Because there is one reversal from left to right at each printing, the original negative must be turned over in the enlarger if an odd generation print is to be chosen for the final image. For our purposes, the first print that you solarize may be taken as the "zeroth" generation, or the "original." The first print from that (the paper negative) is then the first generation print, and so on. The original is seldom chosen as the final result because (1) it is sometimes dark and muddy, (2) it sometimes needs modification, and (3) it is best retained in the files in order to make future prints. The paper negative is most often also retained in the files because it contains all the modifications and is the father of later prints. Creative photography does not demand fidelity to an original image, but only imagination, skill, and a bit of audacity in developing an idea. The final picture may have a full range of tones, be white on black, black on white, or look like a line drawing or etching. This and the following chapters will illustrate how these effects can be achieved, using solarization in combination with the paper-negative process.

One of the most useful accessories in a photographer's workshop is a light box for viewing transparencies of all kinds. A portrait photographer uses one for retouching negatives, a color worker uses one for sorting slides, and we use one for working on paper negatives and diapositives.

The light box may be as simple or as elaborate as you please, the main consideration being convenience. Some photographers just tape their transparency to a window pane, others place a gooseneck lamp behind a sloping piece of glass and still others lay a piece of plate glass across two tables with a lamp on the floor beneath. Such makeshift methods are acceptable only if you use them very seldom. It is tiring and discouraging to use inadequate equipment for long periods, and it makes hard work out of what should be a pleasant occupation. For only a few dollars and a few hours' work you can build a light box suitable for any purpose. The specifications for the one we use are shown in the plans below (Fig. 2-2). It will handle prints up to 16″ x 20″, and a set of black paper masks adapts it to any size smaller than that. The milk glass (also called flashed opal glass) ensures that the light is uniformly diffused, and the plate glass provides a strong, flat, smooth surface that can support the pressure of your hands while you work. The fluorescent lights are sufficiently bright and are much cooler than incandescent lamps would be. Incidentally, the addition of one small ruby bulb and four 15-watt white incandescent bulbs could allow this box to double as a contact printer, replacing the plate-glass printing frame described in the text. A flat piece of Masonite or pressboard should be used to hold the papers in contact during exposure. The fluorescent and incandescent bulbs should be on separate switches, of course, since they are not to be used at the same time. (A physician's X-ray viewing device makes an excellent light box, if you can find one at the flea market.)

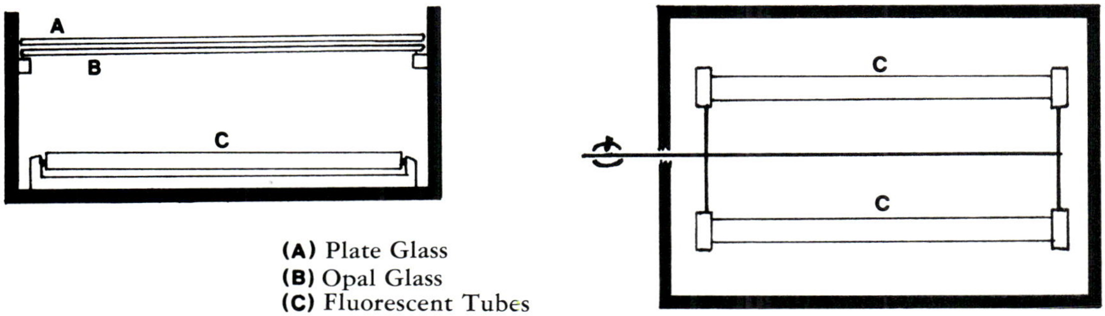

(A) Plate Glass
(B) Opal Glass
(C) Fluorescent Tubes

Figure 2-2. The light box.

Any picture can be used for paper-negative work, but some are more suitable than others. Ordinary scenes that lack atmosphere can be easily improved by the addition of clouds, shadows, or highlights. The suppression of detail or the complete removal of distracting objects is not difficult, and small objects like birds or leaves can be added where needed. Changes in the modeling of a figure can be accomplished, but this must be handled with care lest the figure be made to appear incongruous and unrelated to its background. In such cases it is best to treat the entire picture as a whole, so that the lighting and texture will be compatible with the shadows on the figure. One of the easiest and most useful treatments is to use a soft pencil for shading, increasing the density of selected areas. This is simply a controlled form of dodging, accenting shadows or highlights, when done on the back of the diapositive or paper negative.

The tools you will need are few and readily available. (See Fig. 2-3.) Here is a complete list, and you can actually do a great deal of work with fewer than these.

1. Drawing pencils: 2H, H, HB, 2B, 4B, and 6B.
2. Charcoal pencils: 2B, 4B, and 6B.
3. Acetate drawing ink: black, the kind used for drawing on acetate sheets.
4. Dry tempera color: black, finely milled (or, alternatively, synthetic iron oxide concrete coloring).
5. A roll of surgical cotton and small cotton-tipped sticks.
6. Pumice.
7. Sable paint brushes: one very fine and one medium for detail work.
8. Other paint brushes of assorted sizes for large areas.
9. Kneaded eraser and plastic eraser.
10. Print retouching dyes: Spotone, Kodak, or the like.
11. Spray fixative.
12. Household bleach: Clorox or the like.

The drawing pencils are for detail work where sharp edges are required. The charcoal pencils are for shading and local dodging, since the work can be blended smoothly by rubbing with the cotton sticks. Acetate ink is the most opaque medium you can use. It goes on smoothly with a pen or brush and can be used on either side of the print.

Ordinary india ink is far less satisfactory because it is not opaque and tends to streak when applied to large areas. Dry tempera color or synthetic iron oxide is used for darkening large areas. The density can be built up to any desired degree and the edges softened and blended by rubbing with the pumice on a wad of cotton. Highlights can then be introduced at will by use of the kneaded eraser. The plastic eraser is to be used with the drawing pencils. Art gum is not recommended because it leaves too much debris. The re-

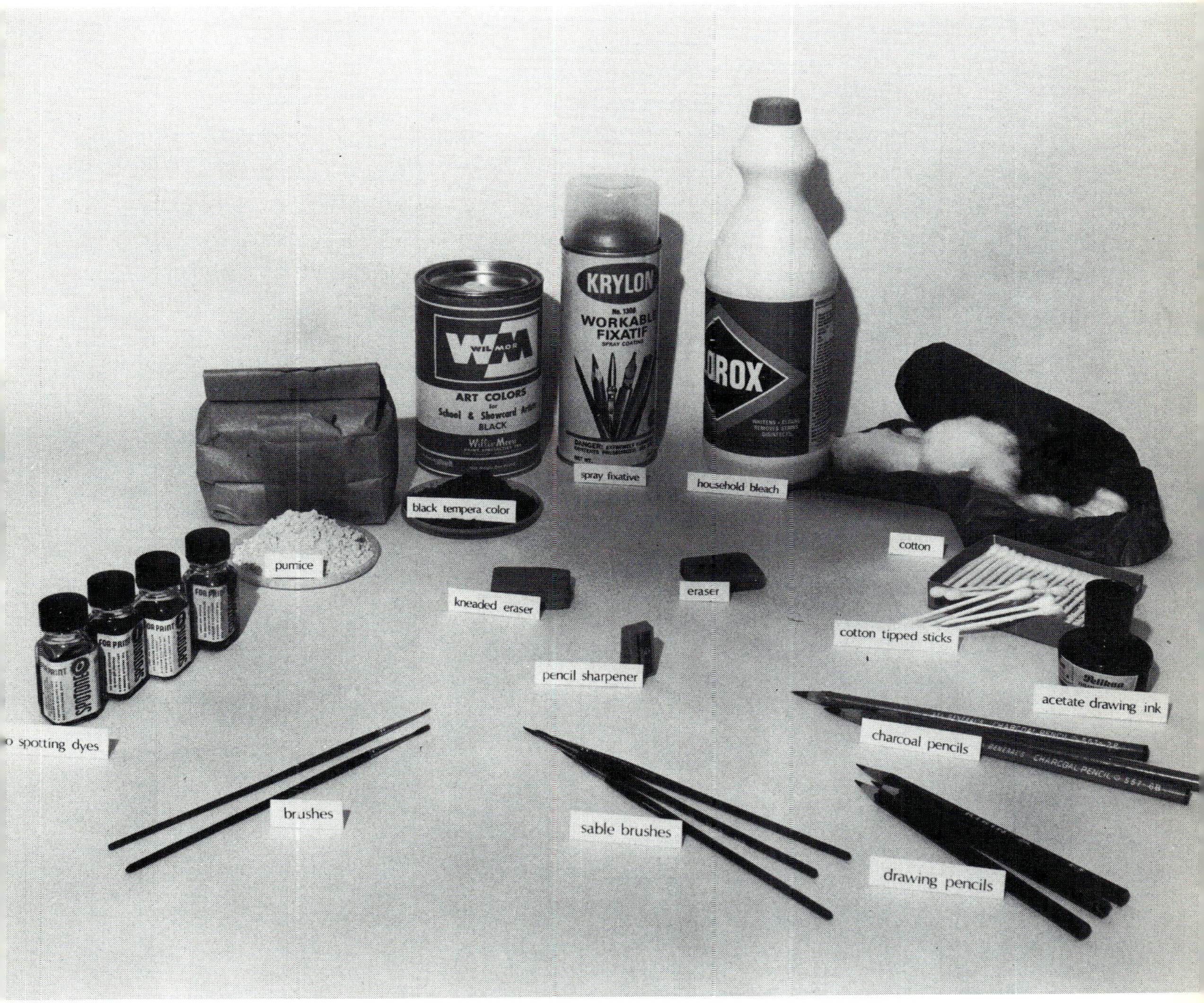

Figure 2-3. Tools for retouching the paper negative and diapositive.

touching dyes are to be applied with a fine sable brush. These dyes come in a variety of shades, and can be mixed to match the tone of any particular print. The mixing is best done with the brush itself in a bottle cap, as only a drop is needed at any one sitting. On comparatively rare occasions, you may wish to remove some detail from a print. This can be done with the household bleach, applied full strength with a small brush and wiped off with a wad of cotton. After a print or paper negative has been thoroughly worked over and is in final form, it should be sprayed with the fixative to protect it from being smeared by handling. If the fixative is of the "workable" variety, it is possible to do further retouching on top of it after it is dry.

All work on a diapositive or a paper negative should be done by transmitted light, and room lights should be dim when using the light box. At first you will find it helpful to check your work by making test prints from time to time. Later you will learn to judge the printing quality by inspection alone.

As you work, you will develop your own particular methods, methods that will be compatible with your temperament and abilities. We illustrate our own methods in this book by showing specific examples of our work. This is only to help you get started. You may find that you can achieve the results you want more easily by some other technique. Above all, if you get an idea that something might work, or just wonder if it would, try it! Even if it doesn't work, you will learn from the experiment, and if it does work—well, life has few satisfactions to match that engendered by a successful experiment.

Figures 2-4A to D. These pictures, on the pages to come, will illustrate a more elaborate procedure, one with increased possibilities for print control, but that requires correspondingly more skill on the part of the photographer. The work is done almost entirely on the face of the print and results in a complete remodeling of the image.

The straight print, 4A, was made on mural finish paper, which has a surface much like that of Portrait Proof R, and is one of the few rough-finish papers easily obtainable in single weight. The print actually treated was slightly lighter than the normal one shown. We see in this normal print that the boat is very cluttered, with a real tangle of stays and lines. The sky is blank and the shoreline dull and uninteresting. The first step in our salvage operation was to darken the entire print by rubbing it all over with black powdered synthetic iron oxide. (As we mentioned before, powdered black tempera color would have done just as well.) The excess powder was then dusted off with a wad of cotton. Next, the kneaded eraser was used on the background land mass to bring it to the same gray tone as the surround area, thus effectively neutralizing it and separating it from the boat. The kneaded eraser was also used to lighten the white parts of the boat, the reflection in the water, and finally to provide some clouds in the sky. To blend the erased areas with the darker parts, powdered pumice was rubbed on lightly with cotton and the excess then blown and dusted off. This softened the outlines and eliminated any signs of the eraser strokes. Following this, an ordinary pencil eraser was used to emphasize the highlights and bring back a little of the original sharpness.

By this time, much of the unwanted clutter and some wanted details were greatly subdued or completely lost. A fiber-tipped pen that draws a fine gray line was used to delineate the rigging, the boat outlines, and the mooring post. Gray was chosen instead of black because this is a predominantly gray print, and black would have been too harsh. As a final step, the corners and edges of the print were given a second dusting with the black powder to produce a slight vignetting, which helps hold the attention on the center of interest. When all of the work had been completed, the print was sprayed with a fixative to protect it against smudging (4B).

From this point, there were two ways of proceeding: We could have rephotographed the print to make a copy negative, or we could have turned the print over and made a paper negative. Since we are now discussing paper negatives, we will naturally deal only with the latter course. (It is also the easier.) The paper negative was made on mural paper also, thus giving double emphasis to the rough texture. It is shown in Figure 4C, and we can see that the lower right-hand corner is not dark enough. A light application of black powder to the back of the negative corrects this, however, and the final print was then made on mural paper (4D).

It is a soft, gray, picture with no clear details. The identity of the boat has been removed, but not its character. It is at rest in a calm, safe harbor, but its tall mast and square rigging remind one of the adventure of travel before the sea winds. This is a mood picture, timeless and placeless, with all visual elements subordinated to the atmosphere of placid calm. Is it still a photograph? Does it matter?

2-4A. The straight print. An adequate but by no means spectacular picture of a boat. A contrastier paper than this No. 2 would give it more drama, but it lacks mostly some suggestion of a sky, and needs to be separated from the background strip.

2-4B. The handworked print. This is the same print as 4A but has the addition of pencil and other work.

2-4C. The negative. Although this print looks very dark, it has all the details required, and does not need another powder treatment. The bottom right-hand corner needed local darkening, so this was done on the back, with powdered black, on a light box. The back was preferable as it will print diffusely and not as a texture. As this paper is also mural, the texture doubled up, creating a much stronger pattern: each paper texture is carried on to the next stage. This can be overused and the image can be broken up too far, but a moderate amount of texture enhances the canvas effect.

2-4D. The final print. Printed on normal contrast mural paper, this print has a charcoal drawing effect. The boat has been separated from the background, clouds have been added, and it only needs toning to look very like a bromoil. Much of the quality is derived from use of the mural textured paper—the powder settles in the hollows and accentuates them. Also, there is some loss of sharpness, as the patterned surface prevents close contact of the emulsions when printing.

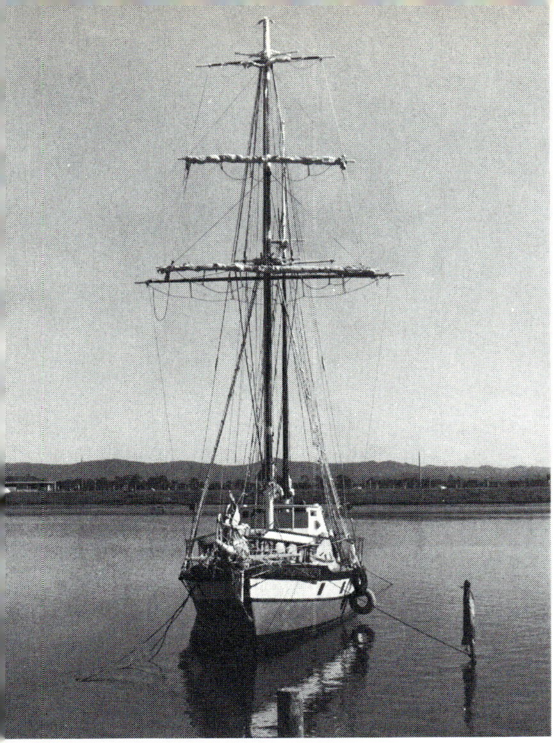

2-4A

2-4C

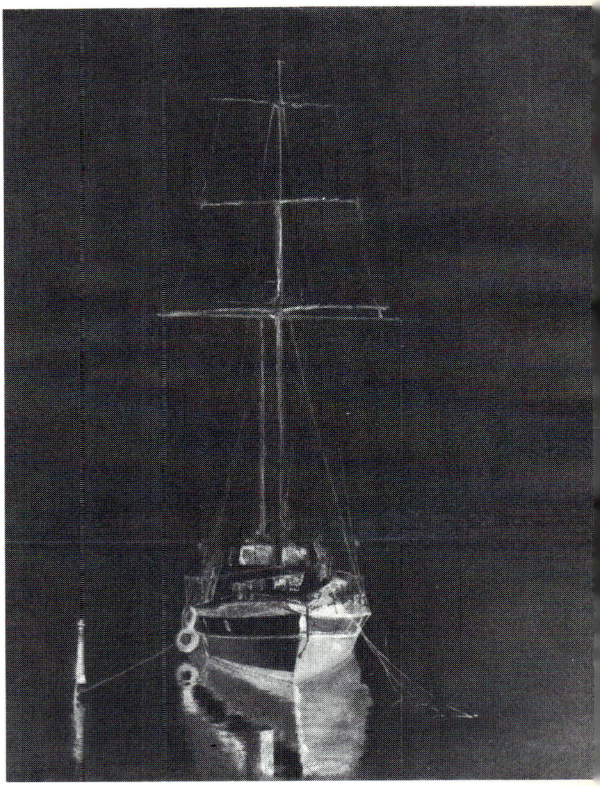

2-4B

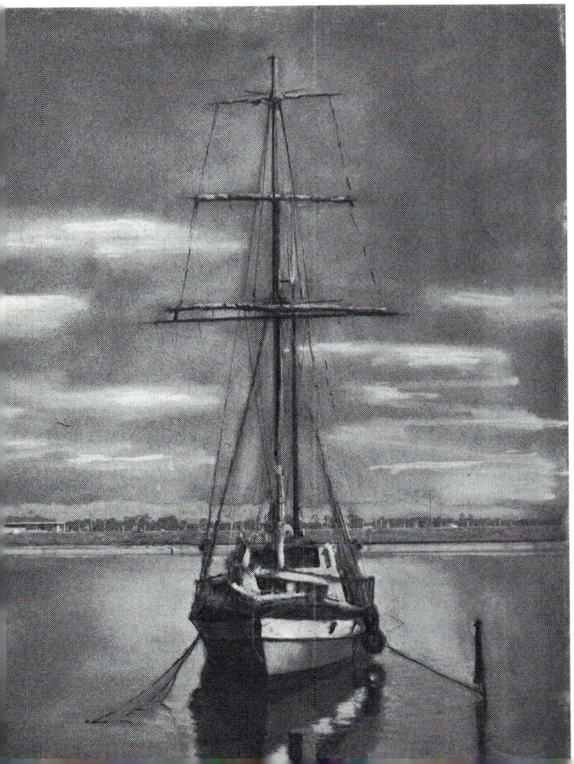

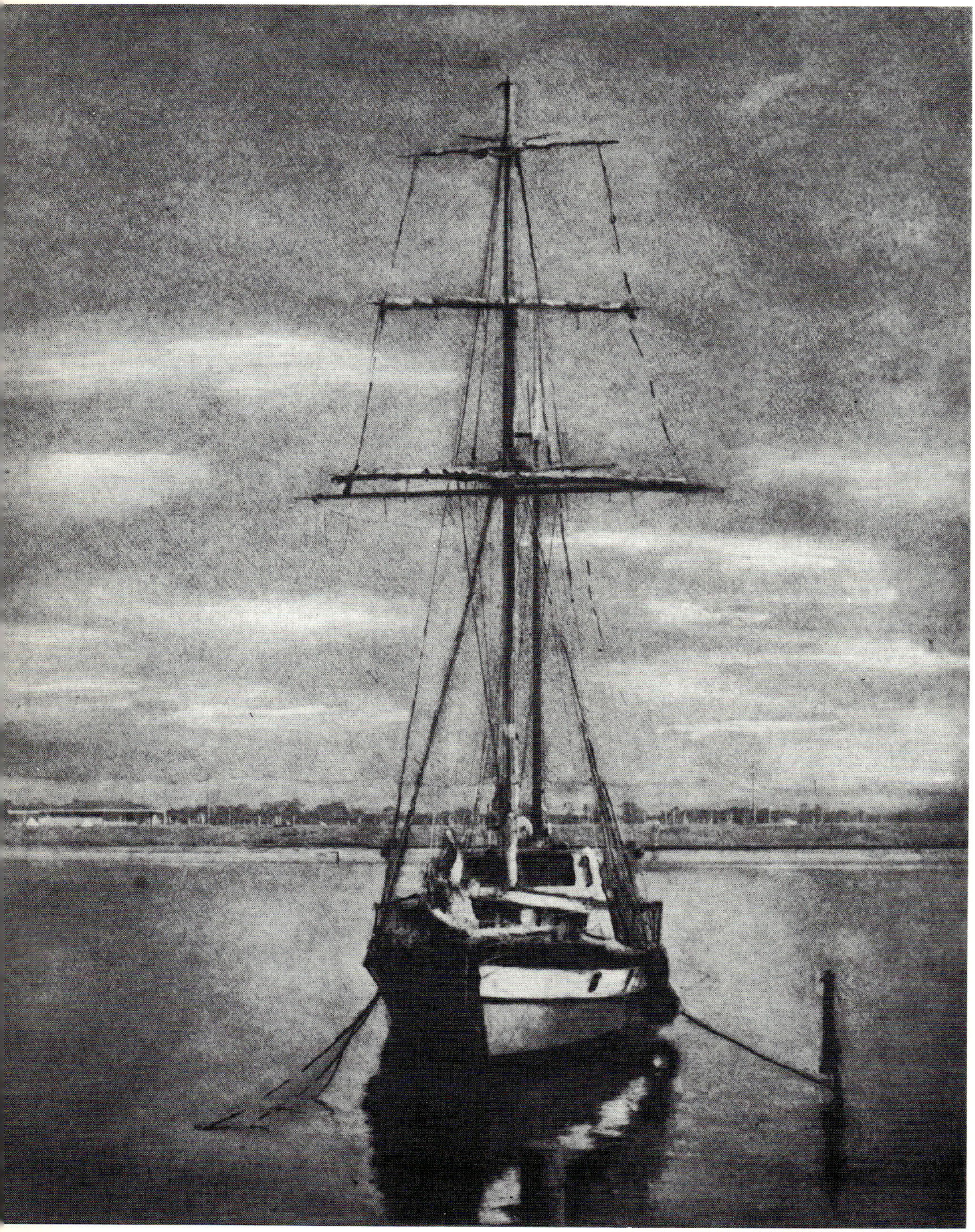

2-4D

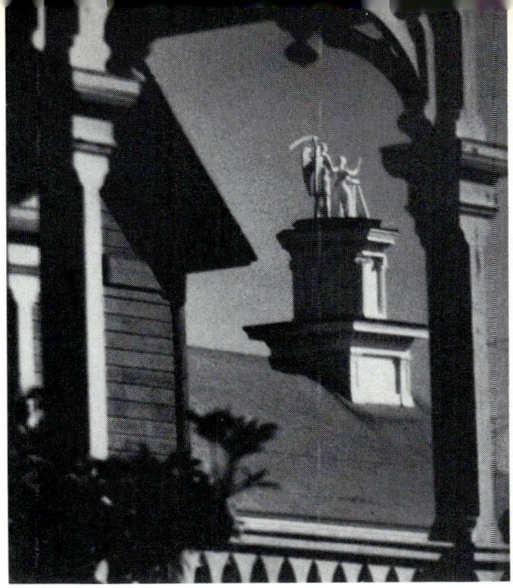

2-5A

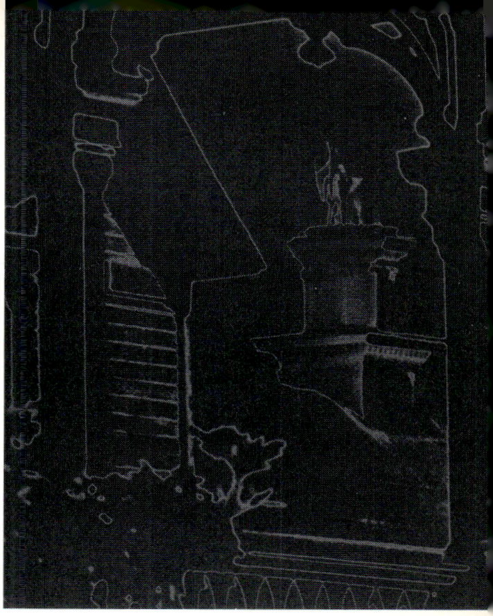

2-5B

Figures 2-5A to G. Statues on the tower. The figures framed by the veranda in Figure 5A stand atop a white building in Mendocino, California. This picture was taken on high contrast film with a long lens on a 35mm camera. The print is made from a small portion of the negative, and printed on a soft paper. It is very grainy, and the image is blurred by camera shake. Figure 5B shows the effect of solarizing on No. 6 paper. The Mackie lines are predominant, but the figure outlines are incomplete, blending with the background. A print taken from this is shown in Figure 5C. It is made on a smooth-finish portrait proof paper, for the intention is to redraw the missing parts of the figures, using acetate drawing ink on the face of the print. Some work with a charcoal pencil—a very soft one, 6B, was used and the charcoal rubbed out evenly with a cotton stick—was also done on the back in order to restore some of the shading for a better balance in the picture (5D). The results of this work are shown in Figure 5E. The final print, 5F, is then made on No. 6 paper, and shows the figures boldly in white against a black background. The Mackie lines have sharpened up the outlines of the veranda, thus removing the objectionable out-of-focus appearance. The charcoal shading on the back of the paper negative appears in the final print as brilliant highlights, giving a feeling of space and depth to the picture. Without it, the print would have been a simple line drawing, a pattern on a flat plane. If so desired, a black-on-white version (5G) can be printed from 5F.

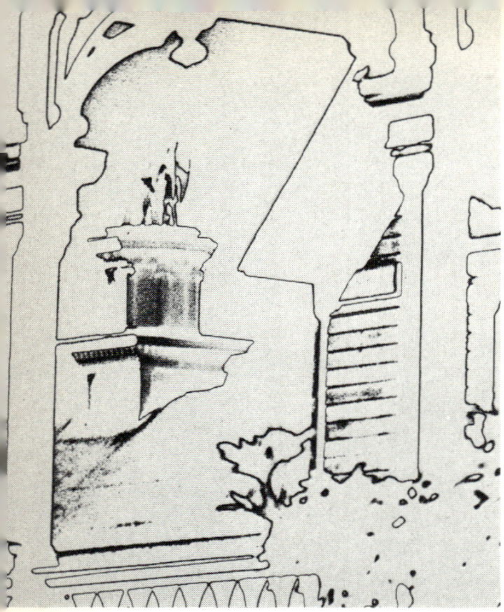

2-5C

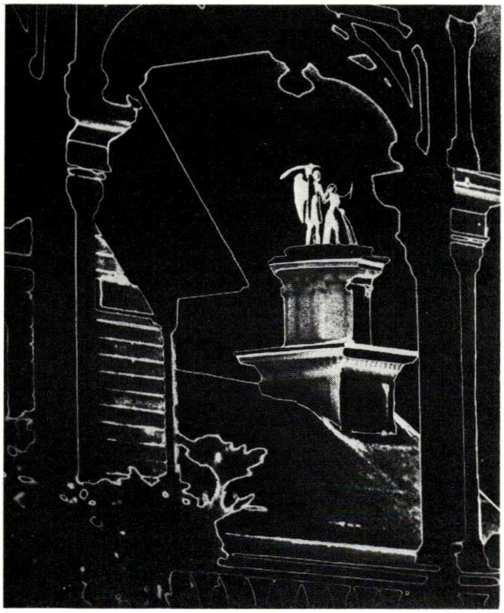

2-5F

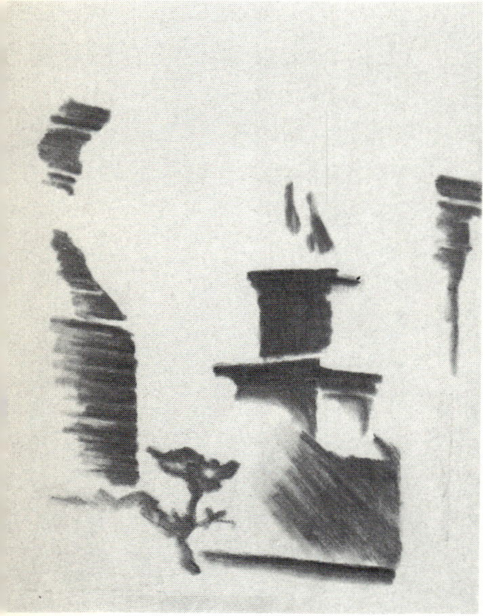

2-5D

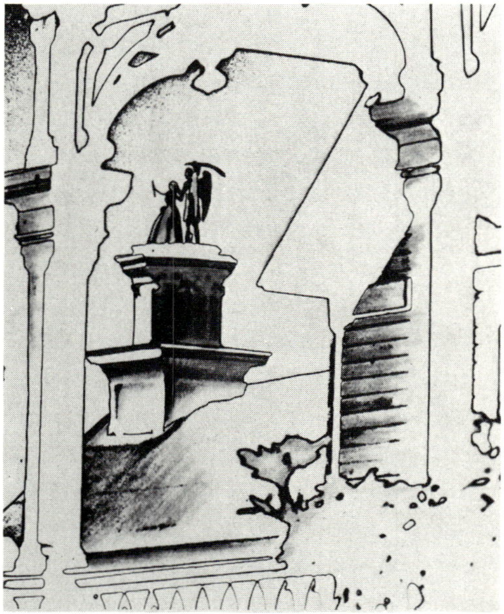

2-5G

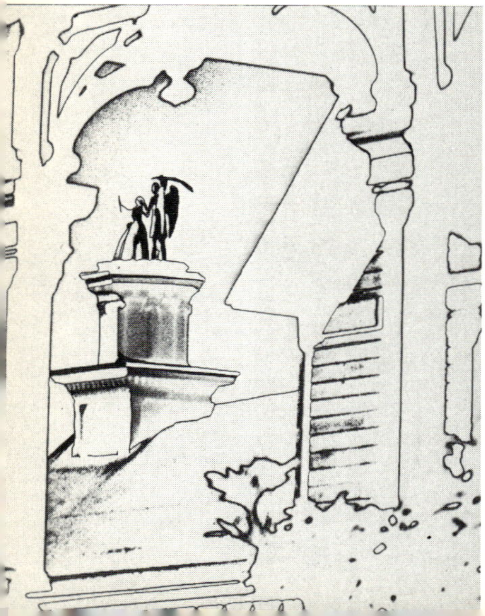

2-5E

Figures 2-6A to G. The statue of the man and boy with the wine press (following page) stands in Golden Gate Park in San Francisco. The straight print, 6A, is unacceptable for two reasons: first, it is a very ordinary picture, deriving its entire interest from the subject matter—that is, it is only a "record shot"; second, the depth of field was insufficient for the subject and the man's face and upper torso are out of focus. The sculptor has already provided us with a strong image, however, so our only task here is to provide emphasis by a more dramatic presentation. A print solarized about midway in the development (6B) shows reversal only of the strongest highlights, and edge effects are minimal. The first generation print (6C) from the solarized original has the major highlights back in key but the heavy shadows have turned white and there are fewer middle tones. A solarized version of 6C is shown as print 6D. It is almost a line drawing in white-on-black, with both the highlights and shadows replaced by their outlines. This print was turned over to make 6E, giving a black-on-white drawing with a suggestion of middle tones. It was felt that this image was too weak, the massive torso being too thinly represented by the reversed Mackie lines. Print 6E was placed face down on the light box and the middle tones darkened with a charcoal pencil. This was a rather easy task—coloring in the outlined areas—almost as in a child's coloring book (6F). The charcoal pencil goes on well and allows full control of the modeling tones, but in this case no attempt was made to blend the charcoal by rubbing, since the coarse texture was desired for the final image. Because the handwork was done on the back of the print, sufficient diffusion would be introduced by the light passing through the paper fibers. While some liberties can be taken with the placement of the highlights in the interest of impact, one must be careful that the illumination of the figure does not appear too illogical. Print 6G shows the result of our handwork. This is a predominantly black print and it is difficult to tell if it is a positive or a negative. We see, however, that the heavy shadows under the right buttock and under the left arm and leg have been retained as black shadow. Notice that the left side of the face is also in its original shadow, but strongly emphasized. The highlights on the side of the body, however, are also black, and the shadow in back of the right thigh is now a highlight.

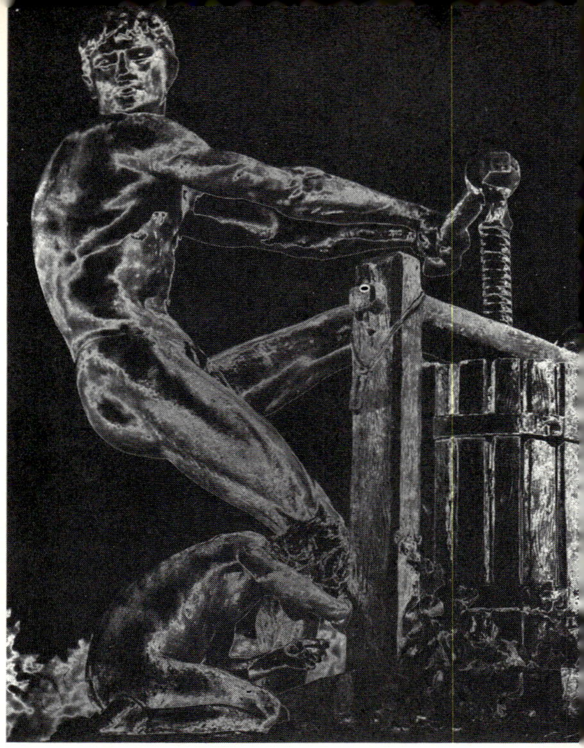

2-6B

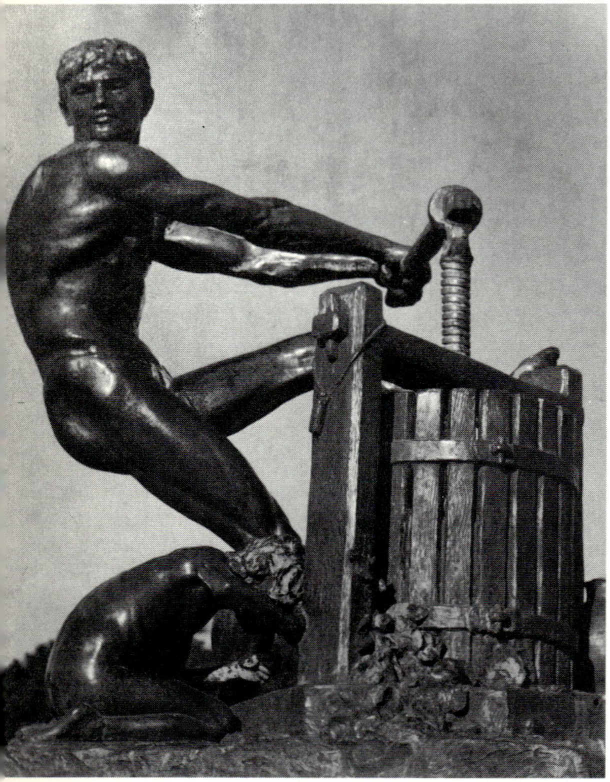

2-6A

2-6C

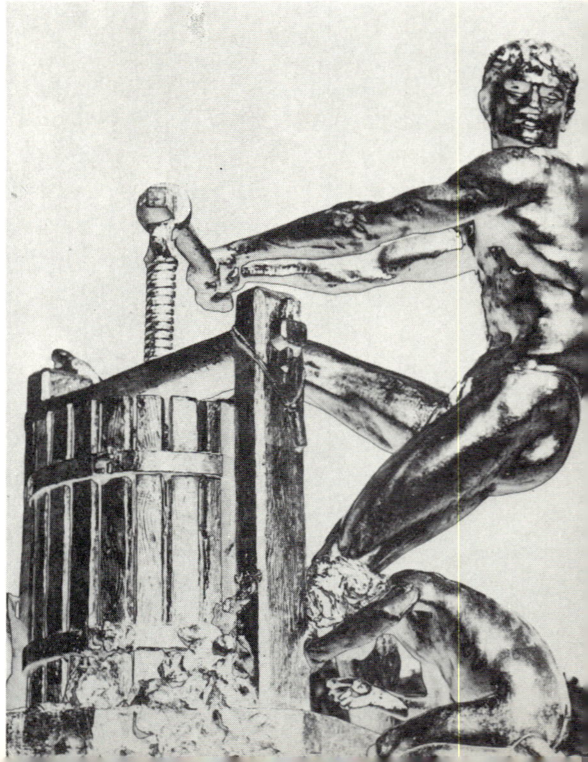

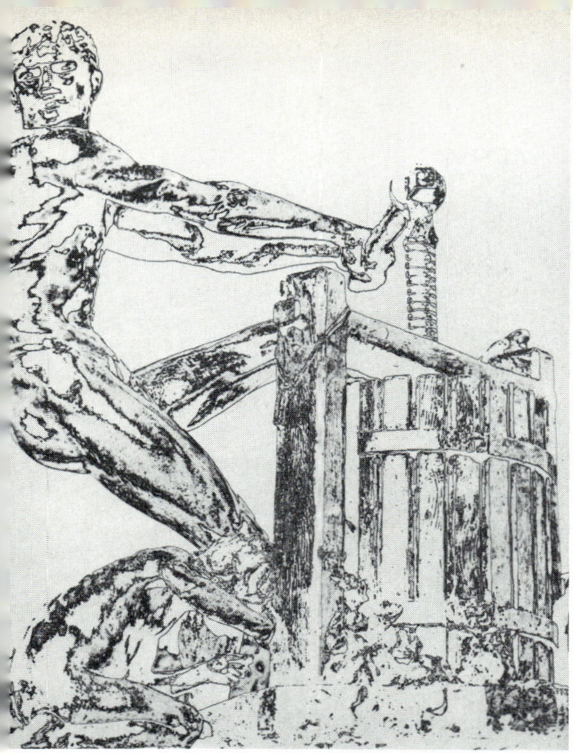

2-6E

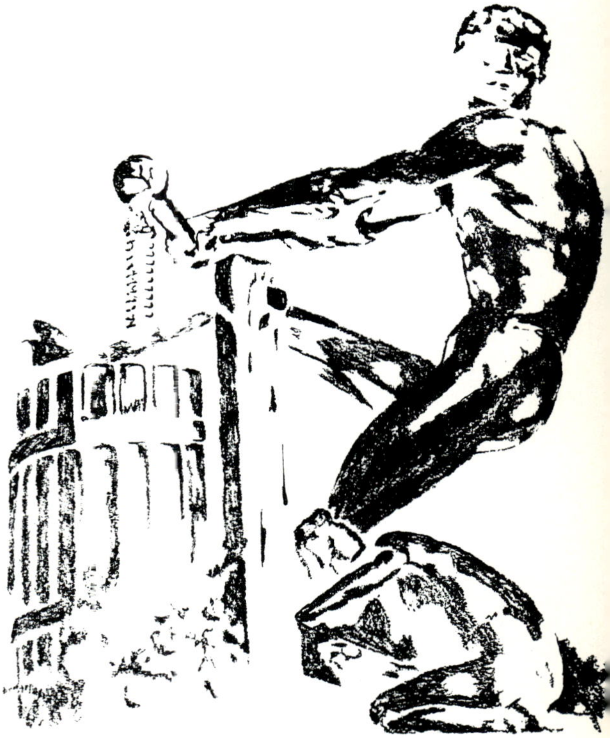

2-6F

2-6D

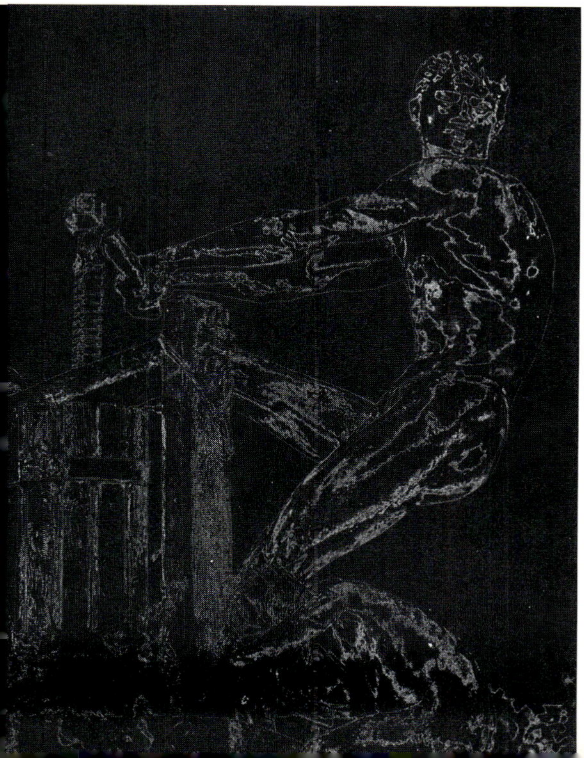

The final print, 6H, which is taken from 6G is the fourth generation and therefore unreversed. It could be considered a positive, although many of the tones are the negatives of the originals. Either print 6G or 6H could be chosen as the final result, and we are not sure which we prefer. The image has been so stylized (or posterized) that the choice would depend upon whether you like a black print or a white one.

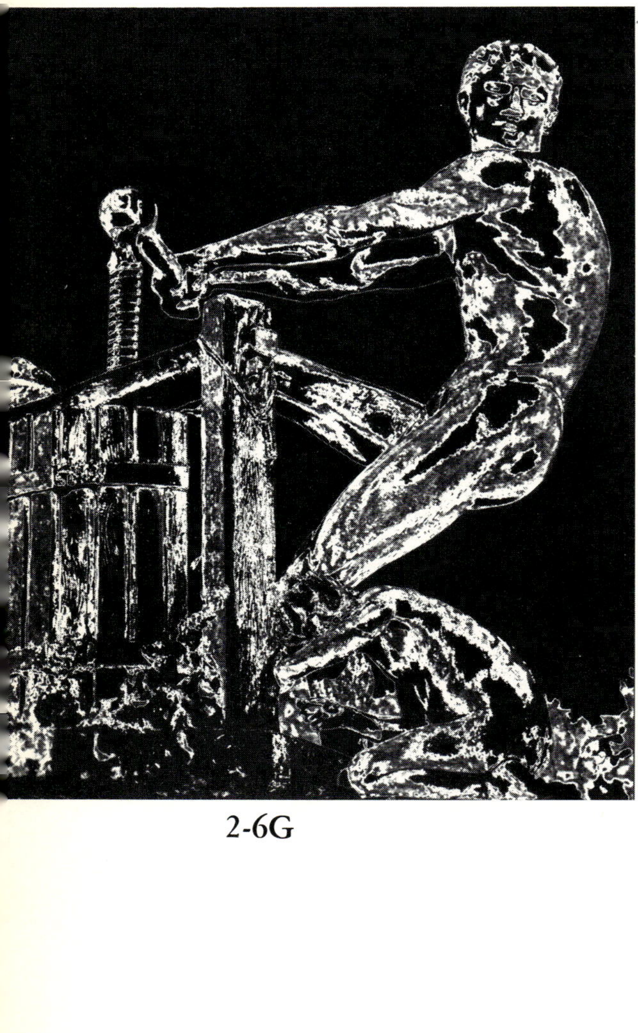

2-6G

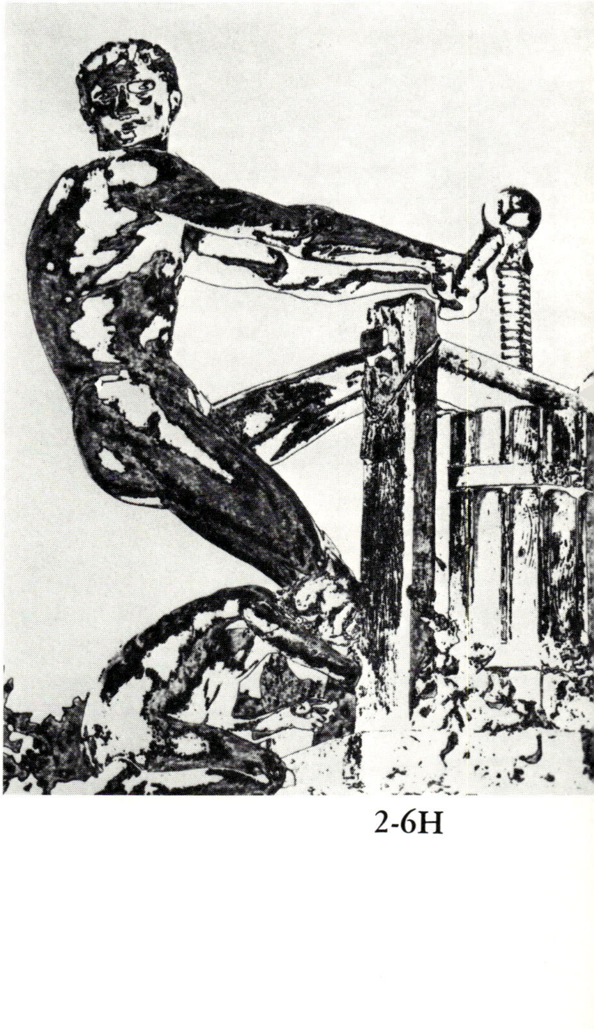

2-6H

3
Special Effects

NEGATIVE SANDWICHES

Perhaps the easiest way to create unusual effects that add interest and meaning to your pictures is to superimpose or juxtapose two or more images. This can be done directly by making a double exposure in the camera, but the results of this method are all too often disappointing. One major problem is that the two subjects you wish to use are not usually available at the same time and place, which means that you must carry one exposure on the film until you get the chance to add the second one, a very inefficient use of your picture-taking time, to say the least. Another drawback is that the placement of the two exposures on the film is very hard to control. Both of these difficulties are minimized, however, if your camera has a ground-glass viewing screen and removable film holders. Then you have only to mark the position of the first subject on the ground glass and preserve the film in its holder while you wait for the chance to take the second subject. One good thing about this double-exposure method is that the density of the developed image responds approximately to the logarithm of the exposure rather than to the exposure directly. Thus, putting two exposures on top of one another will not double the density of the negative, but only increase it as much as a one-stop overexposure. If each subject is underexposed by one stop the maximum density of the negative will be no more than normal. The tonal integration of the two exposures is usually quite good under such conditions, and the negatives are not difficult to print. Several images can be put on the same negative if they are photographed against a black background and not allowed to overlap too much. This kind of work, of course, is best done in the studio where the lighting can be carefully controlled.

Another method of making composites of images is the "negative sandwich" technique. The images are on separate negatives, which are sandwiched together and placed in the enlarger as one unit. In this way most of the subject difficulties of the double-exposure method are avoided. Not only can you choose subjects taken at widely different times and places, but the negatives can be oriented as you please, and even turned back-to-front if necessary. A square format is the most satisfactory from the standpoint of versatility in orientation, and large negatives are much easier to work with than small ones.

The biggest problem of the negative-sandwich method is contrast control. Unlike the double-exposure process, here the densities *do* add together, and where two dense parts of the negative overlap, the combined density is often unprintable. As an example, consider the imaginary case of two identical negatives sandwiched together in register. Let us suppose that in each negative the highlight areas transmit 1/4 of the enlarger light, while the shadow areas transmit 3/4 of the light. This is a contrast ratio of 3 to 1 between the illumination of the shadows and highlights on the print. When the two negatives are used together, however, the highlight transmission is only 1/4 of 1/4, or 1/16, of the incident light, while the shadow transmission is 3/4 of 3/4, or 9/16, of the incident light. This is an illumination ratio of 9 to 1, or three times that of the single negative. Thus it frequently happens that when two negatives of average density and contrast are sandwiched together, there will be some parts of the combination that are very dense indeed, and will show up only as white areas on the print. To avoid this difficulty, the chosen negatives should be quite thin—exposed normally, but underdeveloped by about 30 percent. A normal negative often works quite well with a very thin one, especially when the thin negative carries only a texture or a pattern.

Successful negative sandwiches usually require that one of the images be taken against a dark background so that the surrounding film is almost empty, allowing the other negative to dominate the background. A simple rule of thumb: If you are going to combine images by sandwiching negatives, shoot against a black background; if you are going to combine them by double-printing (see below), shoot against a white background. Good combinations of images often can be found by laying out a set of negatives on the light box and trying various pairs in different orientations. You can greatly increase the number of such good combinations, however, if you keep their possibilities in mind while you are photographing. It is a good practice to make two negatives of every likely subject, exposing one normally and the other one stop less. Or, if both are exposed normally, the development time of one should be normal and the other one third less. The most satisfying combinations are those that have been previsualized before the negatives were exposed. After some experience, you will find yourself thinking in terms of superimposed or juxtaposed images, and be able to adjust the image placement and exposures to match the effect you wish to achieve (Fig. 3-1).

Negative sandwiches that suffer from the extreme contrast problem described above still can be used effectively in some cases if the resulting print is solarized. The solarization reduces the contrast and fuses the two images together, producing more of an integration than a superimposition of the two. Symmetrical images like the one in Fig. 3-2 are easily produced from two identical negatives sandwiched face-to-face. The variety of such combinations is almost endless, and they have a kaleidoscope-like fascination.

Figures 3-1 & 3-2. The simplest way to begin the exploration of fantasy in photography is to make a negative sandwich. Just place two negatives together in the enlarger carrier. The proper choice and combination of negatives are critical to good results, but not difficult to achieve. Thin, low contrast negatives are required to ensure that the combination is printable, since the densities of the negatives are added and the contrast is greatly increased. Randomly selected negatives will often reveal surprising images when

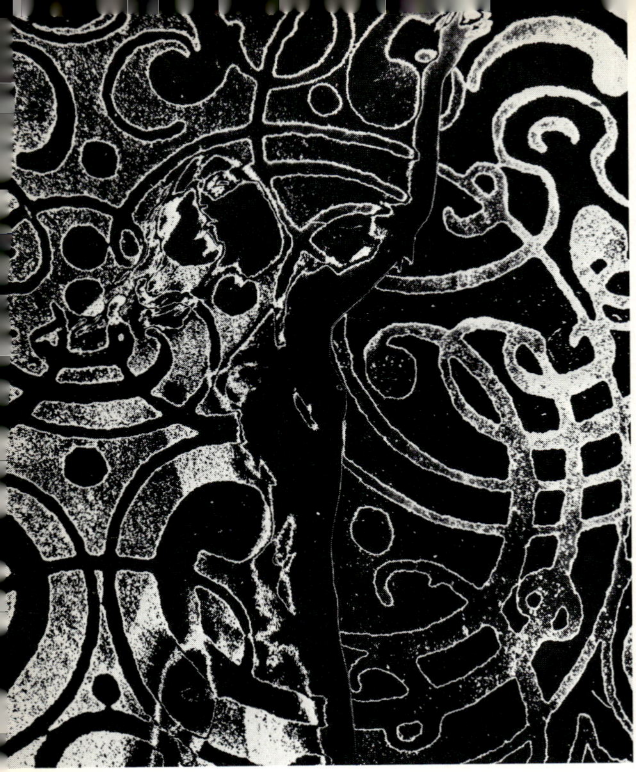

3-1

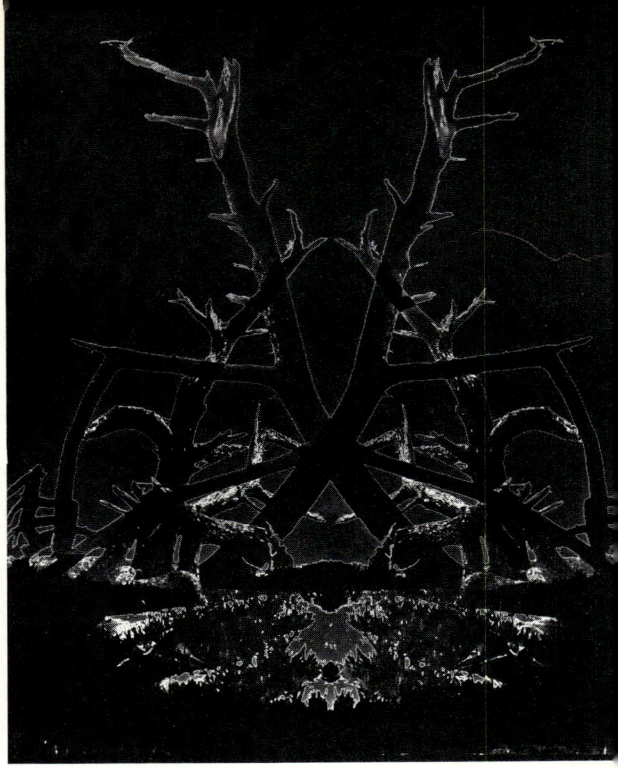

3-2

superimposed, and it is a fascinating experience to play with a set of negatives on a light box, discovering the enormous number of patterns that can be made by their combinations. Many of these may appear exciting to the eye, but be quite disappointing when printed, because printing paper cannot handle the excessive contrast. Also, you will find that many of the images are not suitable for juxtaposition, most often because they are unrelated and their combination would be pointless. But, if the images have any meaning for you, as they must if you are a photographer, you will soon begin to understand the possibilities and limitations of the sandwich method, and be able to choose negatives that will combine with printable and significant results. It is quite likely that you will soon be taking pictures with the idea of superimposing them and adjusting your exposures to that end. If you have a negative that is suitable except for its density or contrast, don't overlook the possibility of making a copy with the desired characteristics. A positive transparency, either in black-and-white or color, may sometimes make an effective combination with a negative.

The print made from a negative sandwich can be further removed from realism by solarization. This also has the effect of integrating the images so that the method of combination is undetectable. Even in those cases where the negatives are badly mismatched and the print is unbalanced due to excessive contrast, it is often possible to achieve an acceptable fusion. The solarization fills in the unwanted white areas and at the same time draws a delicate line about the images, effectively tying them together. This has a tendency to flatten the perspective, reducing the apparent depth of the scene, giving, in extreme cases, a poster-like effect.

DOUBLE PRINTING

The first darkroom "trick" that a photographer learns is to print more than one image on the same sheet of paper. This is certainly the simplest way to add clouds to a bald sky, and almost anyone can do this successfully with very little practice. In many cases, the technique is carried no further than this, for one soon finds that more complex images are not nearly so easy to combine by double printing without leaving telltale edges that give away the secret. To add the clouds, one simply shields the rest of the print while exposing the sky region to the cloud negative. The cloud image is so light and so diffuse in character that a slight overlap is of no consequence, and is undetectable if the exposures are properly matched. Adding a figure to a landscape or combining two strong images from different negatives, however, is another matter. There are two separate problems: achieving compatible exposures, and blending the edges at the joining boundary. The first of these can be handled by choosing the negatives carefully, and by making test strips to determine correct exposure. Negatives taken under very different lighting conditions do not blend very well, nor do negatives of very different contrast. A predominantly dark scene is not going to be easily matched with a high-key image which is mostly white. The selection of compatible images is one of the primary exercises of your creative ability, and if you handle this selection successfully, you will not likely have much trouble choosing negatives to match. The second problem, that of blending the edges so that the joining will be invisible, is one of technique, which can be easily learned. The secret is in the proper masking of the successive images, and registering the masks accurately over the print.

A mask is simply a shield, shaped to cover the part of the image that you do not want to print and reveal the one that you do. Masking can be done in several ways, and only experience can tell you which method you will find most convenient and effective for your particular application. Some photographers apply the mask directly to the negative in the form of an opaque retouching medium. A number of commercial preparations are made for this purpose, including some that may be either washed off or peeled off after use. Of course, the application requires some skill, and it is still necessary to dodge the region of overlap to obscure the effect of the hard edge. To avoid the danger of spoiling a valuable negative, it is often advisable to do the opaquing on a separate piece of clear film, which is then sandwiched with the negative in printing.

The method that we prefer, because we find it simpler and more versatile, is to use cardboard cutouts on a masking stage. The stage itself is nothing more than a piece of plate glass, somewhat larger than the print, and two 4″ x 4″ x 8″ blocks of wood. The blocks are positioned on either side of the printing easel and support the glass four inches above the surface so that the printing paper can be easily slid in and out. The first negative is placed in the enlarger and focused through the glass onto the easel. A piece of cardboard, larger than the print, is placed on the glass stage, and, without refocusing, the outline of the desired image is traced on the cardboard. The image on the cardboard will be smaller than the print image and slightly out-of-focus, but still sharp enough to trace with sufficient accuracy. The cardboard is then removed from the stage and laid on a piece of Masonite or some other firm cutting board, and the traced outline is carefully cut through with a sharp razor blade or etching knife. This cut must be made cleanly, without whittling, because both the two pieces are now put back onto the glass stage and fitted together, the cut-out image being exactly in register with the projected image, as it was when traced.

Wide masking tape is used to anchor the outer edges of the masks to the glass and to form hinges so that each piece may be folded back and replaced in exactly the same position. After determining the correct exposures for the two negatives by test strips, the first negative is placed in the enlarger and carefully registered in the correct position with the mask. The printing paper is slid under the glass and positioned in the easel, and the first exposure made with only the first half of the mask in place. With the safelights still on, the second negative is substituted for the first, the first half of the mask and the printing paper are removed, and the second image is positioned and focused on the easel. Then the printing paper is replaced in the same position on the easel, the second half of the mask is put in place, and the second exposure is made. If the work has been carefully done, the two images should be blended without telltale evidence of a dividing line. In most cases, however, it will be found necessary to do a little dodging at the edges of the masks in order to hide the joining line.

The above procedure turns out to be more difficult than it sounds, and actually requires a good bit of practice before successful blends can be made with confidence. Often, there will be a narrow region which is either too dark or too light, or an unwanted bit of overlap of the two images where the mask was not quite accurate. Now, if the purpose of the composition is to make a "fake" photograph, that is, one which looks like a straight photograph of the composite scene, the only course is to try again with more care and skill. But if the aim is an artistic composition rather than a realistic one, and the picture is to be solarized, we do not have to worry about any defects of blending for they will be completely submerged and negated by the complex of changes and reversals that occur during solarization, and no one can tell whether the picture required one negative or a dozen.

There is nothing sacred in the four-inch height of the masking stage just described. The glass could be laid directly on top of the printing paper, or raised almost to the enlarger lens. Some workers put the masks in the filter holder just below the lens, and some just hold the masks in their hands and keep them in motion like dodging tools during the exposures to assure a soft edge. The closer the mask is to the printing surfaces the sharper will be the projected edge. The nature of the images to be blended determines the sharpness needed, so it is not wise to be too rigid about the height of the masking stage. The four inches suggested above is a compromise height, suitable for many

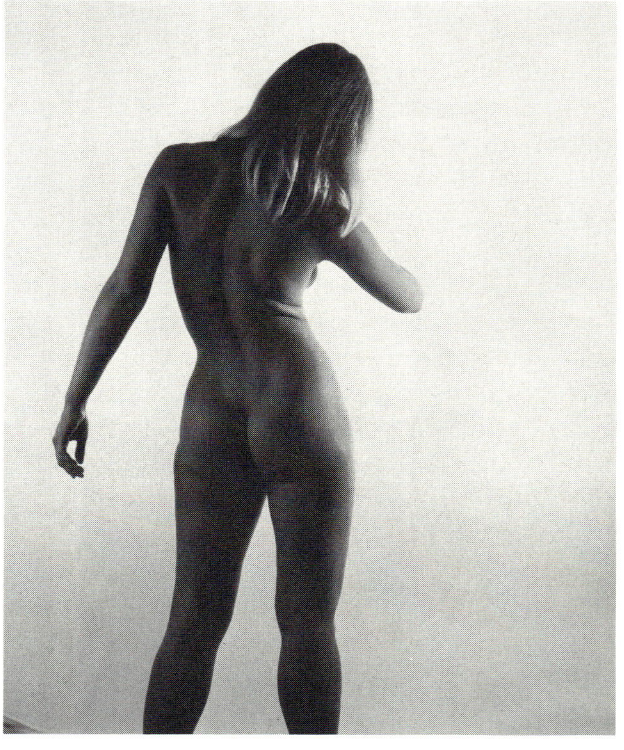

3-3A

3-3B

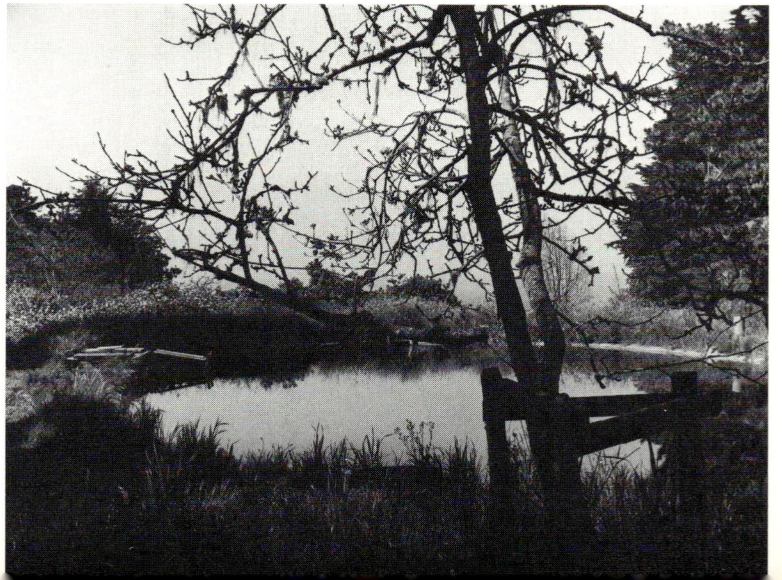

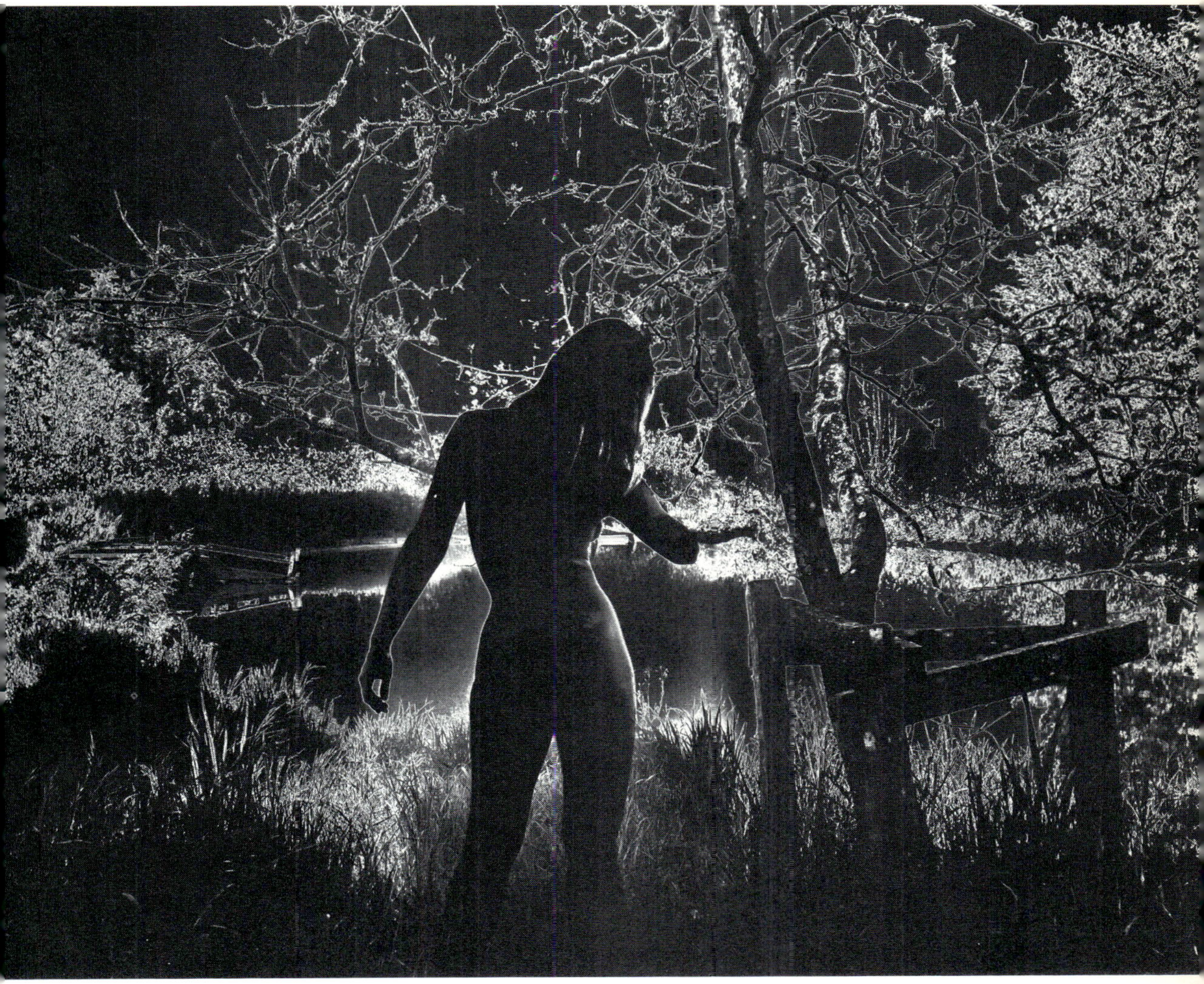

3-3C

Figures 3-3A to C. Two pictures (3A and 3B) were combined by double printing. The millpond was printed first, while a mask reserved a spot for the figure. The scale of the enlargement was changed so that the girl would fit naturally into the setting. After both negatives had been exposed, the print was developed about one minute before solarizing to blend the two images into one mysterious night scene (3C).

subjects, but certainly not for all. Also, the printing paper must be protected while the second image is being focused and positioned. Since there is a danger of spoiling the registration whenever the paper is moved, it is often safer to focus through the red filter of the enlarger or to simply cover the printing paper with a piece of mounting board.

When the boundary between the two images is a straight line the problem of joining them is greatly simplified. In this case there is a well-known technique that is easy and effective, whether the composition is to be made by a double exposure in the camera, by a negative sandwich, or by double printing: Simply mask off the unwanted part of the image when taking the picture. This is done by holding a card blocking part of the lens or using an opaque half-disc in a filter holder at a certain distance from the camera. There are two things to remember. First, the light rays striking the top part of the film come from the bottom part of the scene, those striking the right half of the film come from the left half of the scene, and vice versa. Second, the farther the mask is held from the lens, the sharper will be the boundary. A single-lens reflex camera or one with a ground-glass viewing screen will allow you to see exactly the effect of the mask, so neither of these considerations should give you any trouble. The result of blocking half of the lens is a negative that is half clear film and half picture, with a gradual change from exposed to unexposed part.

The combination can be made in the camera by blocking the other half of the lens and making the second exposure on the previously unexposed area of the same film. Getting the right combination of masking and exposure can require a little practice or experiment, and some sort of adjustable shield (in a filter holder) is usually called for in order to insure good results. A better procedure, however, is to make only one exposure on the film, and to make the composition in the darkroom, using the negative-sandwich method. The negative with one half exposed and the other half clear film is ideal for combination with another of a similar kind. Here, there is no trouble with overlapping images or increasing contrast because of addition of densities. The gradual decrease in density of one negative can be made to just compensate the gradual increase in density of the other negative so that the joint will always be smooth.

With a little more effort, properly masked copy negatives can be made in the darkroom, so that foresight is not needed in taking the original. In fact, the extra amount of control in choosing and matching the images may make this the preferred practice for many photographers. It offers the possibility of combining images from any of the pictures already in your files, no matter for what purpose they were originally made. Negatives made from color slides can be combined with images from black-and-white prints and solarized images can be combined with unsolarized ones or positive ones with negative ones. There is no substitute for experimentation. The more things you try, the more you will learn, even if the experiment doesn't work.

REPEATED IMAGES

The repetition of a single image often has a fascination beyond that of the image by itself. A section of wallpaper or a sheet of postage stamps presents a visual experience quite different from the flower pot or presidential portrait, because the orderly design subordinates the single image. In Figs. 3-4A to D we show two versions of a repeated image, one a symmetrical, self-sufficient design of two elements, and the other an open design of endless repetition. Each began with a solarization of the same print (Fig. 3-4A), a girl digging in the sand. She appears in strong white outline, but her reflection has gone negative in Fig. 3-4B. Before solarizing, the negative was printed twice on the paper, once in reverse. The symmetrical image (Fig. 3-4C) is simply a print from the solarized original. The positive and negative versions are equally acceptable, depending on your choice. The repeated pattern (Fig. 3-4D) required a mask with the cut-out image taped over the opening. The printing paper was moved according to a measured pattern to produce the twelve images by successive exposure.

Figures 3-4A to D. One variation of repeats is to use the solarized paper negatives to produce the same image twice, with one opposite or upside down. In Figure 4C we have taken the girl digger and her reflection (4A) and faced them opposite each other. This was done by twice exposing in the enlarger, first with the negative emulsion down, as normal, then second with the emulsion up, to reverse her, printing each time on only one half of the paper while masking the other half from stray light. As the reflection on the wet sand was lighter than the true image, it solarized only softly compared to the outlined solarization of the girl herself (4B). When this was reprinted again as a paper negative, as in 4C, the strange effect of identical diggers with identical shadows was made more strange by the fact that the shadows were much heavier.

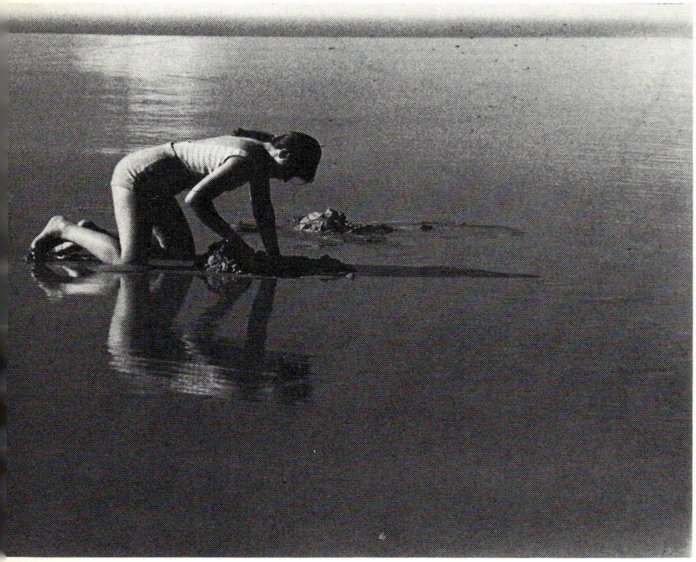

3-4A

3-4B

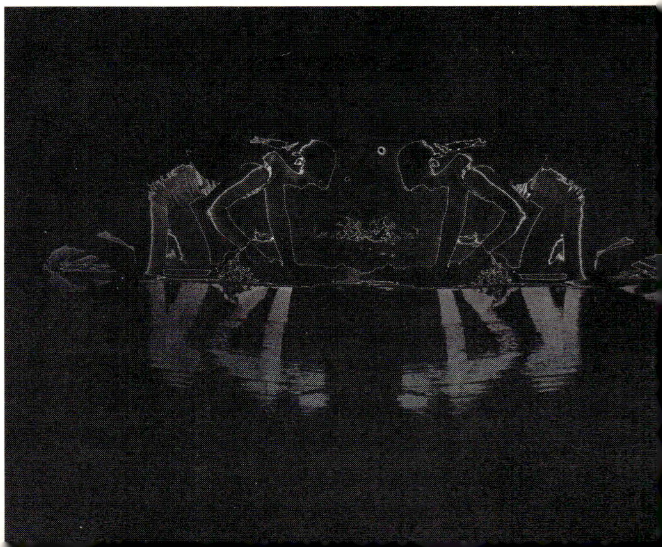

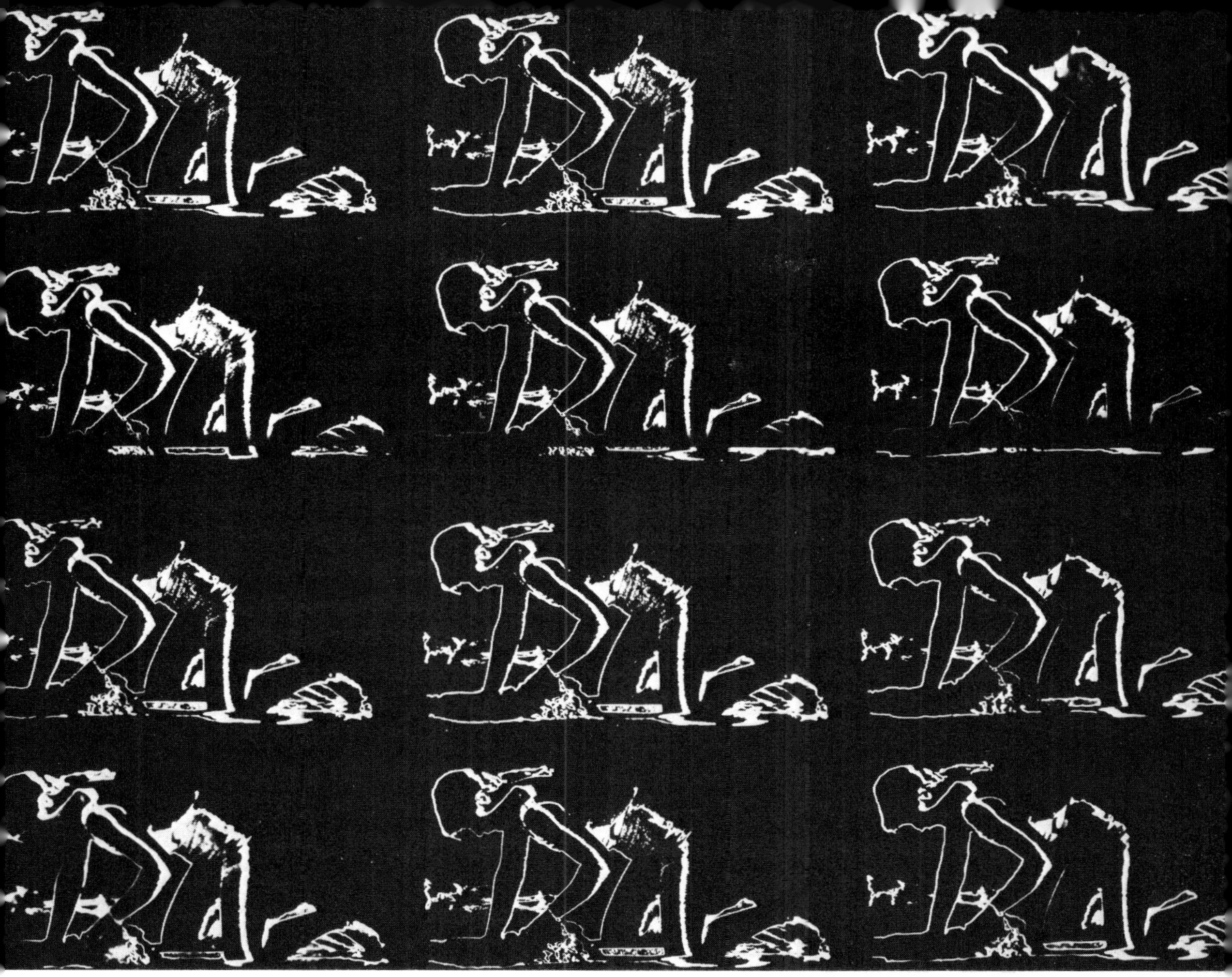

3-4D

Another variation of image repetition is shown in Figure 4D. This time the same image was used to create a repeated pattern. The cut-out image was taped over the opening in a mask and the printing paper was moved in a measured sequence, producing the twelve images by successive exposure.

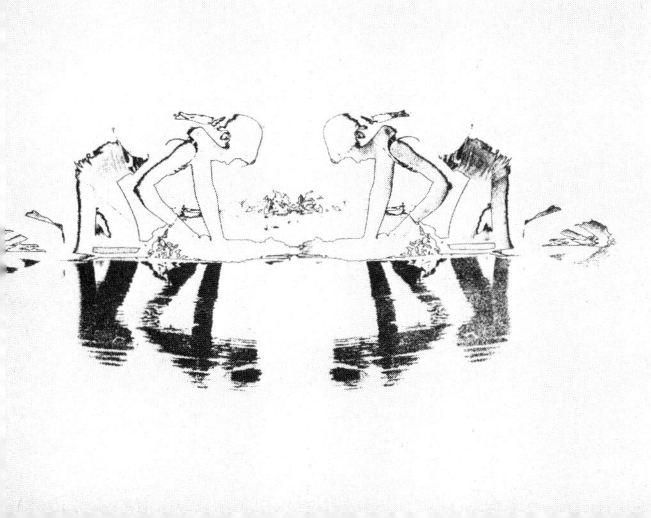

3-4C

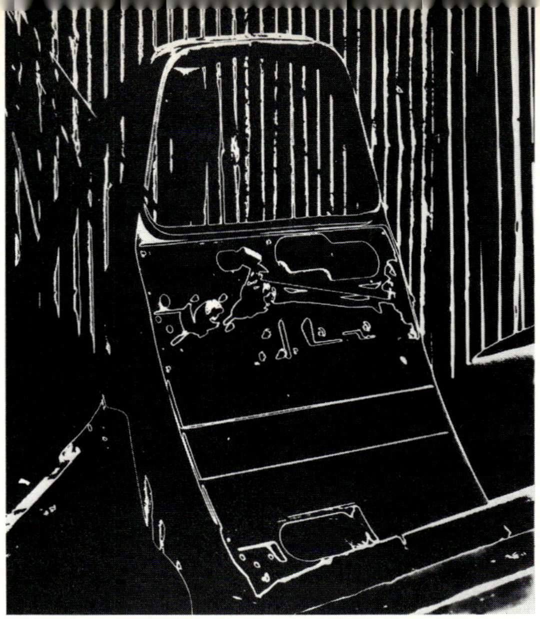

3-5

LINE DRAWINGS

One of the major objections to straight photographs as works of art is their tendency to show too much. Without trying to define what is and what is not visual art, we can still safely say that two of its characteristics are selection and emphasis. A photographic record of a scene that shows all objects with equal value is not usually considered artistic, but one that, by selective focus, lighting, or perspective, concentrates your attention on one object and emphasizes some significant feature or aspect of that object may well qualify.

An idea of feeling can often be communicated more strongly by a few lines than by a full-scale photograph. Only an outline is needed—the imagination of the viewer will do the rest, the image in the mind's eye having the strength of a personal meaning. In our pictures, the real, the factual, and the particular have been abstracted, leaving only the idea, design, or the symbol; and the purest form of this abstraction is the line drawing produced by solarization. (See Fig. 3-5.) The process is simple and straightforward if the original negative is of high contrast. For a low contrast negative intermediate steps must be taken to increase the contrast. To illustrate the method, we start with the highest contrast possible—a shadowgram or photogram. Lay an opaque object on a piece of printing paper under the enlarger with no negative in the carrier. Expose the paper sufficiently for it to go black when developed, leaving a white shadow of the opaque object. When the paper is black, turn on a strong solarizing white light and watch until the white shadow begins to darken. Turn off the white light and continue the development for at least one minute longer, until the shadow is as dark as the rest of the paper. Complete the processing—the paper will look completely black by reflected light, but if you hold it up to a light and look through it, you will see a very fine white outline of the object you placed on the paper. This paper negative can then be printed to give a fine black outline of the object, reproducing all of the edge detail far more accurately than any artist can trace it.

In many discussions of the Sabattier effect, you are told that it is most effective pictorially when used with simple outlines and large areas of solid masses. Here, we find that just the opposite is true. The more detail and fancy filigree there is in the original, the more interesting and beautiful the final design. In a line drawing, large blank areas tend to be boring and allow the picture to "fall apart" unless the surrounding line is particularly intricate or interesting. A blank sky in a landscape will remain blank when solarized, but when the sky area is framed with tree branches, the solarized pattern of leaves and branches will effectively keep the interest within the picture. Our shadowgram of the clamp (Fig. 3-6) is shown only as a pedestrian example of a process, but its pictorial effect is much improved by the delicate detail of the screw threads. The picture of the two seagulls and posts (Fig. 3-7) began as two negatives of one seagull and one post, silhouettes against a white sky. The two were printed on the same piece of paper and solarized.

3-6

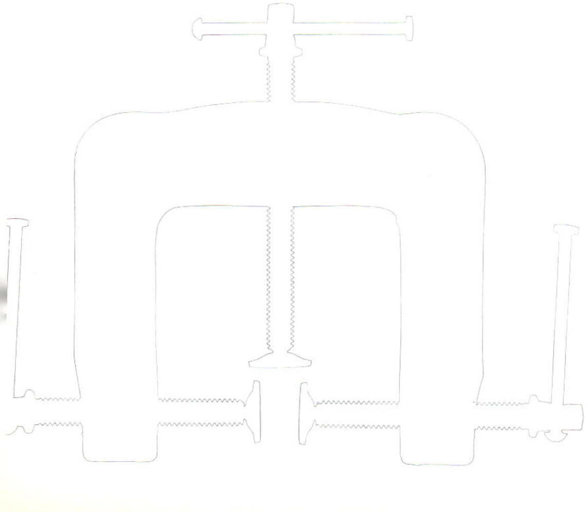

3-7

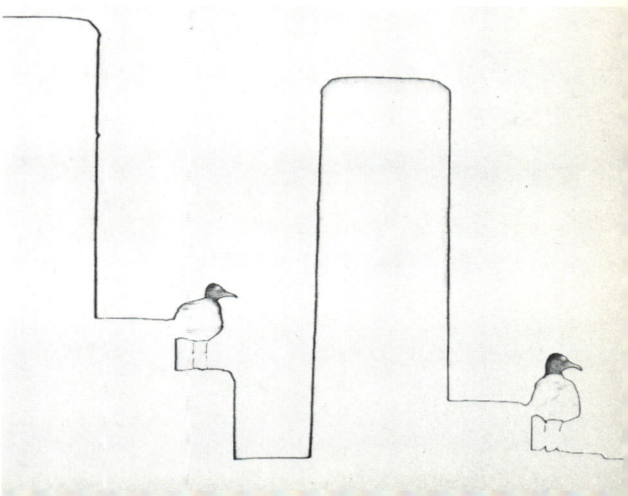

The posts and birds printed black, and the sky went black during solarization, leaving only the Mackie line around the double image. When the solarized picture was turned over and printed, it gave this cartoon-like design of extreme simplicity.

A more complicated example (Figs. 3-8A to E) shows a pile of bricks turned into an abstract design, starting with a slightly contrasty negative. The two figures in the next picture (Fig. 3-9) were photographed in one of the endless corridors of the Los Angeles airport, and this treatment conveys something of the surreal atmosphere of these surroundings. The Monterey cypress tree and the tree branch design (Figs. 3-10 and 3-11) have carried the process one step further. Each of them has been solarized a second time, producing a second Mackie line and giving the picture a more delicate appearance.

Figures 3-8A to E. Line drawings of bricks. Figure 8A is a straight print on No. 2 normal contrast paper, indicating a high contrast negative, with black shadows, taken on a bright sunny day. Figure 8B is the same negative printed on No. 6 paper, eliminating most of the details in the brick surface. Solarizing this print produces Figure 8C, with a fine white Mackie line only, the rest having gone through to black. Turning this print over and printing yields Figure 8D, the black-line drawing on white paper. If this is used as a paper negative it will yield Figure 8E, a white "crayon" drawing on black paper.

3-8A

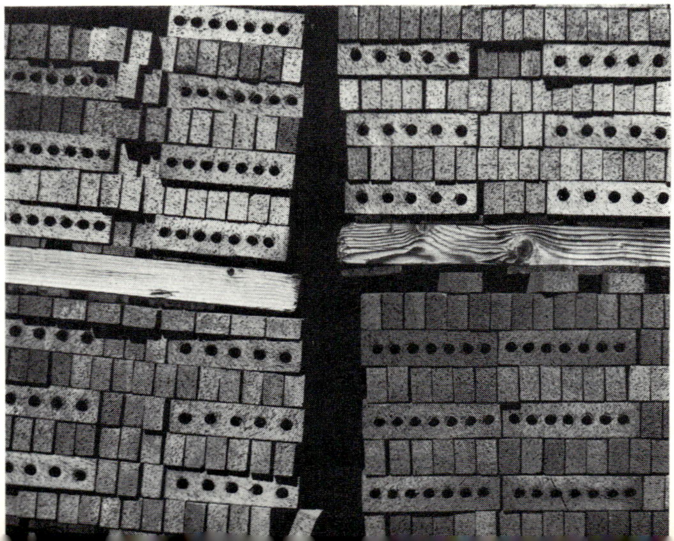

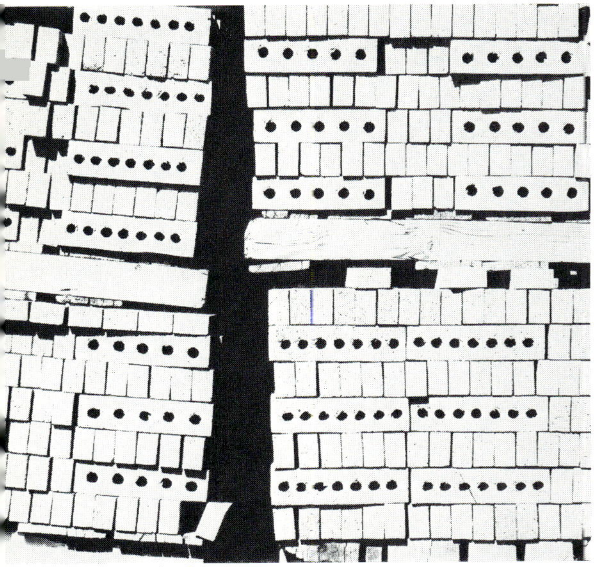

3-8B

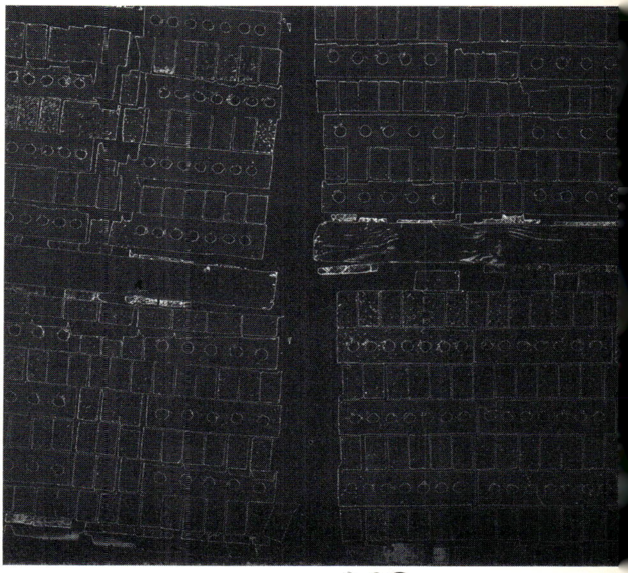

3-8C

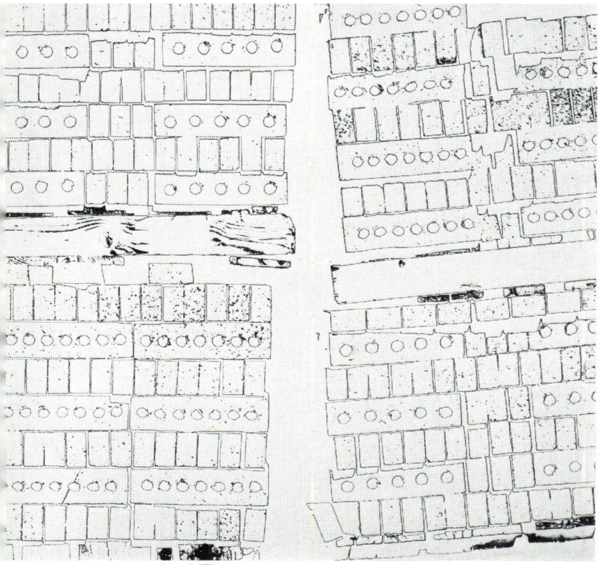

3-8D

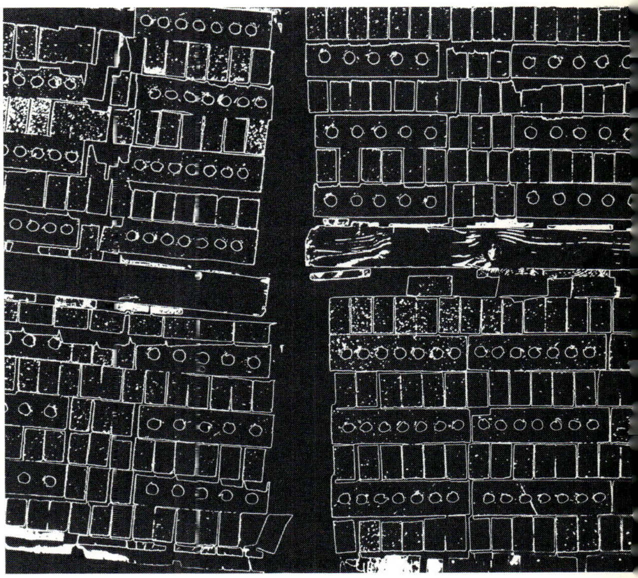

3-8E

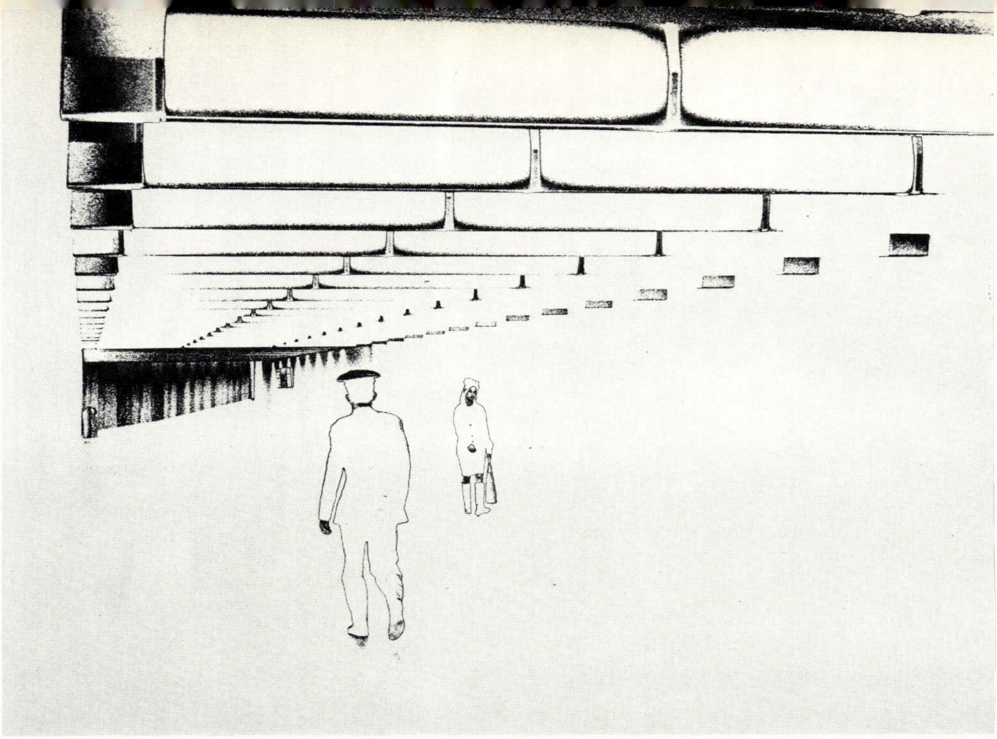

3-9

3-11

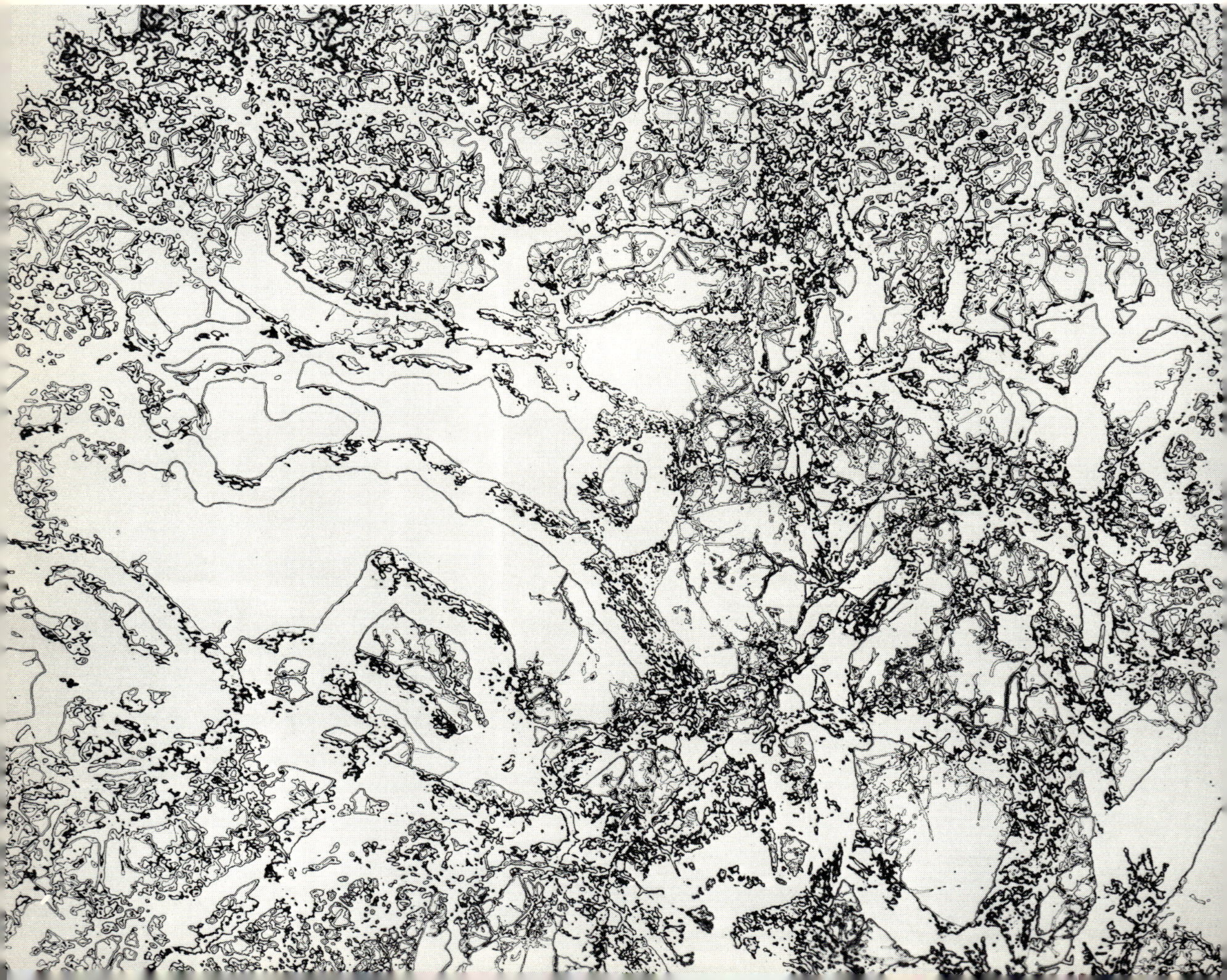

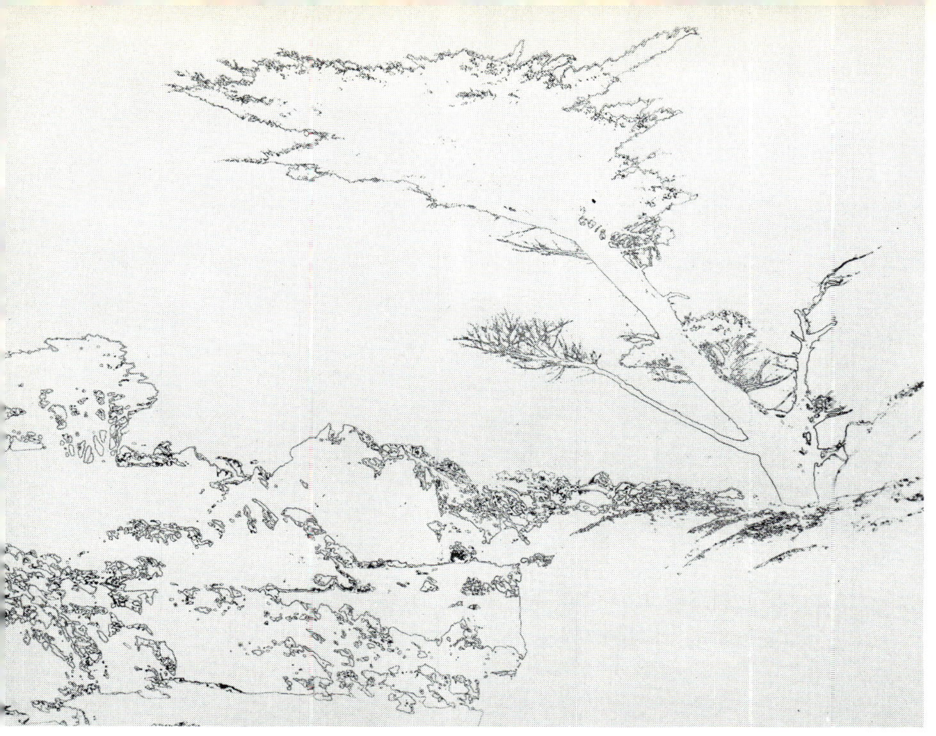

3-10

BLOCK PRINTS

Linoleum block and woodcut prints have a distinctive character because of the continuous smooth printing surface and the all-or-nothing nature of the printing ink. Letterpress and gravure processes use screens to produce halftones by means of many small dots that are not resolved by the eye. Block prints, however, have no halftones, and one made with black ink on white paper is everywhere either black or white and the halftones have to be suggested by cross-hatching or stippling. High contrast photographic materials can imitate this effect if the middle tones are eliminated by repeated printing. Starting with a negative of normal tonal scale, make the original print on No. 6 paper. Turn that over to make the first generation print (a negative), and from that make the second generation print (a positive), and so on. The fourth generation print will be a positive with no middle tones.

Most of the fine details will also be missing, giving a very poster-like image in black and white. Middle tones, but

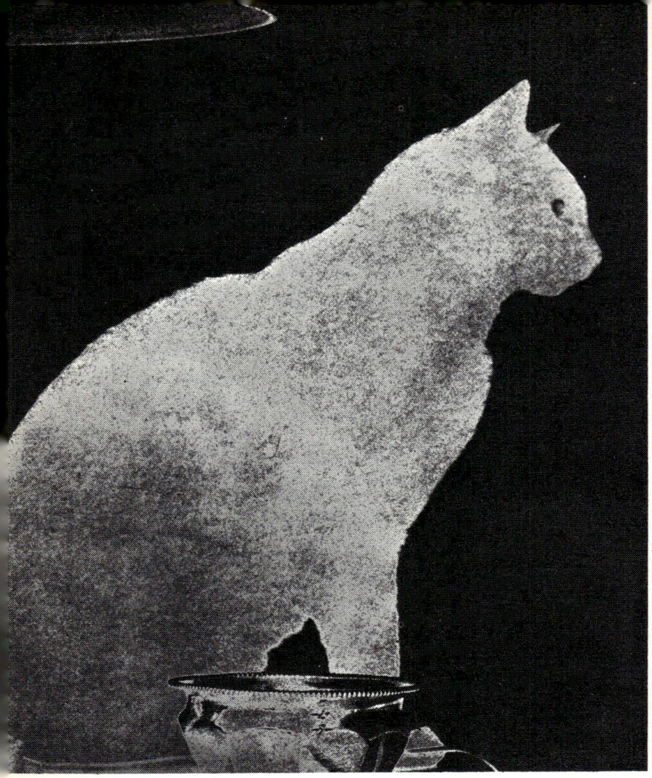

3-12

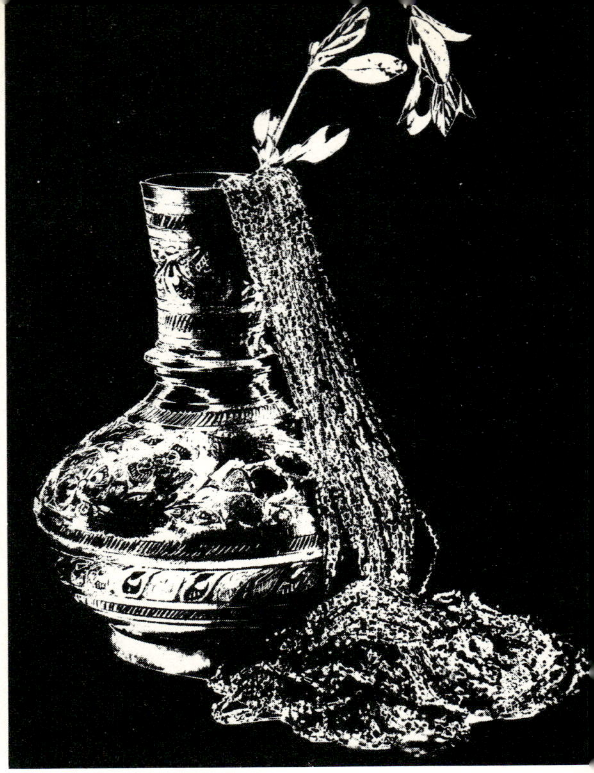

3-13

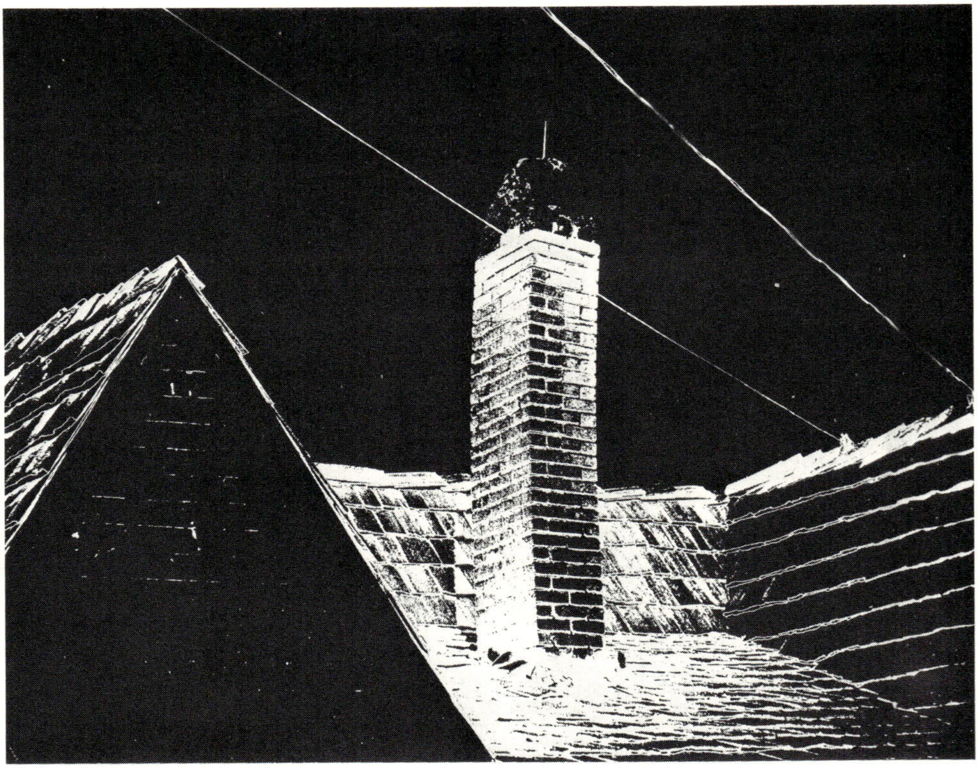

3-14

not the details, can be reintroduced at any stage simply by solarizing. From your experience with the Sabattier effect at this point, you will know how to hold or to produce a grey at will by flashing the print at the proper time during the development. For example, the grey cat in the picture (Fig. 3-12) was printed from a lightly solarized positive, and was developing as a pure white silhouette. When the bowl had reached approximately the grey desired, the white light was turned on and left on until the cat had achieved the same value of grey. The fuzzy texture is not the cat's fur, but the grain of the paper positive showing up. Although the original exposure had not been enough to cause the cat to develop, it was sufficient, when added to the second exposure, to produce the grainy texture.

The pictures of the rooftop (Fig. 3-14) and the vase and twig (Fig. 3-13) were both originally photographed against a white background on high contrast film. The first prints were strongly solarized, and all intermediate tones were lost by the second generation prints shown here. This "block print" effect is so dramatic and easy to achieve that you will see many examples of it, including some in the gallery section of this book.

ETCHINGS AND TEXTURE SCREENS

An etching is made from a copper or zinc plate that has been exposed to the action of a strong acid, yielding an image consisting of many small pits, which hold the ink for transfer to the paper. The shape of the pits depends upon the artist and the tools that he uses to draw the image, and may take the form of lines, scratches, or dots. While it is very easy to imitate this appearance by using a texture screen as shown in Fig. 3-15, in many cases the paper-negative process itself introduces a grainy structure that has the character of an etching. The dancing "Woodnymph" in Fig. 3-16 demonstrates this characteristic. On the other hand, the grain can be deliberately emphasized by printing through the back of the paper, but this also results in a slightly softer image quality. Paper textures vary according to the brand and type

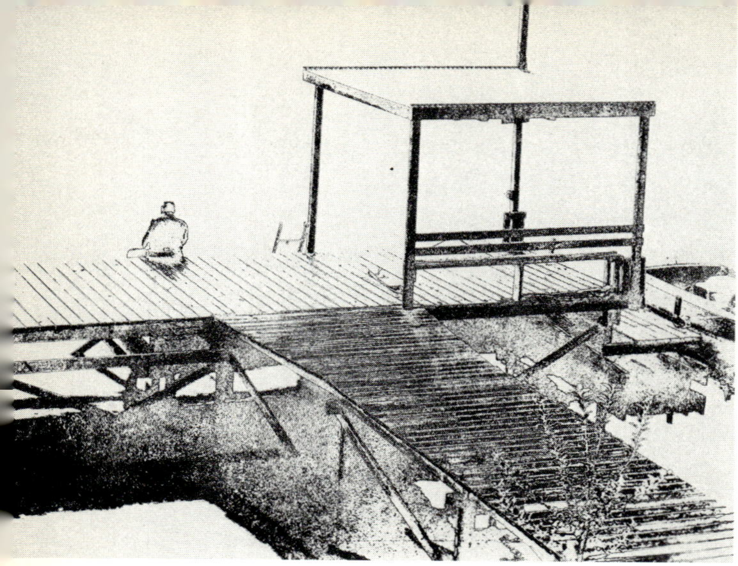

3-15

3-16

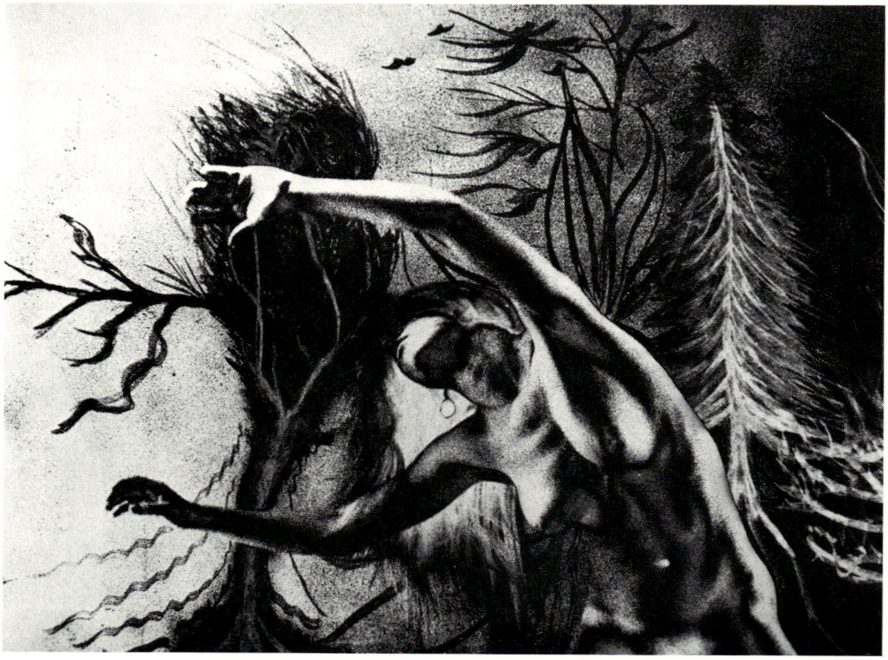

3-17

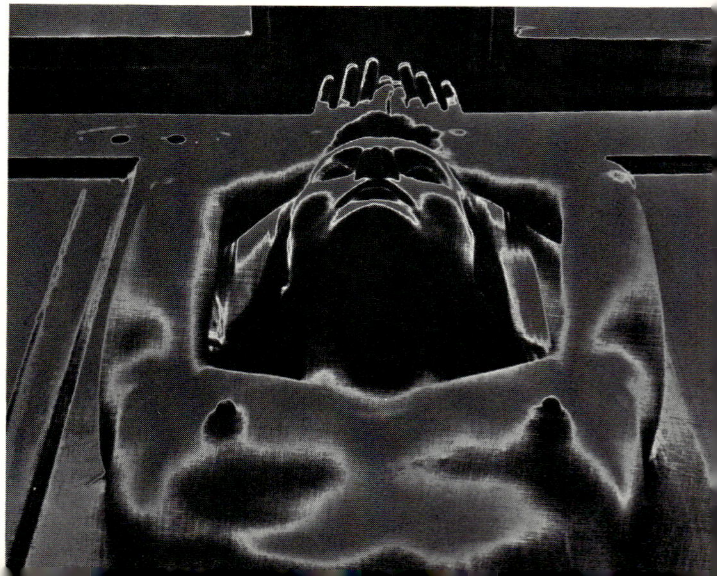

of paper, as you can see by comparing the examples given in Chapter 2.

A texture screen to simulate etchings can be easily made from a sheet of lucite or other clear plastic. Using a very fine sandpaper and even regular strokes, score the plastic with a cross-hatching of scratches, horizontally, vertically, and diagonally. The plastic sheet can be sandwiched with the negative if the lines are fine enough and the negative is large enough. With a coarser texture of lines it is better to have a large sheet of plastic that can be placed over the printing paper on the easel. The result of printing this screen in contact with either the paper or the negative is an image criss-crossed with many fine white lines—the reverse of an etching, which is made up of fine black lines. The white-line image looks somewhat like a linen texture (Fig. 3-17). One might think that the white lines could be made black just by solarizing the print, but this turns out not to be so simple. Even if it were possible to reverse the lines without reversing the image, the Mackie lines, being white, would compromise the result (Fig. 3-18). To get black lines on a solarized print, the print should be solarized first, then printed with the screen to produce a paper negative with white lines, which when printed in turn will give a positive with black lines.

3-18

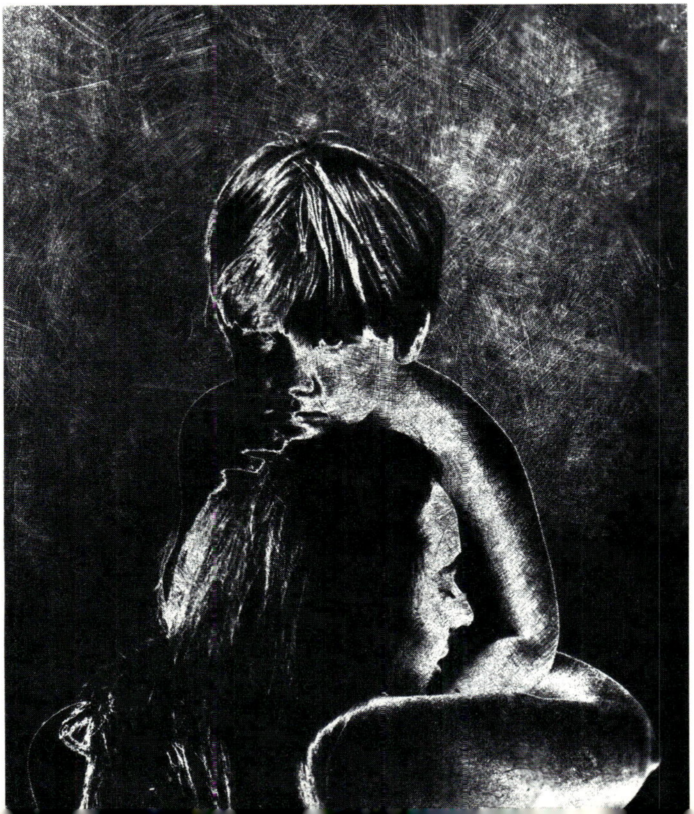

Other Texture Screens

Once you have begun to use texture screens for any purpose, you will quickly discover that there is an almost endless variety of textures, and that almost any texture effect can be reproduced photographically. The simplest texture tools to use are the patterned glass panes that you can buy from the local glazier. He will probably have twenty or more different ones for you to choose from. Buy a piece two inches wider and longer than the prints you plan to make, and have the glazier grind the edges so they will not cut your hands. This is then placed over the printing paper for all or part of the exposure. In some cases, you may want the pattern to be in soft focus while the rest of the image is sharp. To accomplish this, raise the glass a short distance above the printing surface, supported on books of the proper thickness. The dancing torso in Fig. 3-19 was originally printed through such a piece of glass for part of the exposure before being reprinted as a negative and solarized.

Some years ago, in the golden age of pictorial photography, texture screens of many kinds could be purchased from photographic supply houses, and their use was an accepted part of the craft. With the coming of the purist movement in photography, however, the vogue for texture screens passed, and they were scorned by many as symbols of the decadence and artificiality of the pictorial era. They almost vanished from the market for a time, as there was no demand, but, fortunately, have now reappeared. Their inevitable revival was aided by the cultural revolution that took place during the 1960's and brought us rock music and psychedelic art. Texture screens, patterns imprinted on film large enough to lay over the print, are available in styles that imitate different types of cloth, such as canvas, tapestry, or linen. Sizes larger than 8″ x 10″ cost several dollars each, so a varied collection of them could be rather expensive.

Fortunately, new products in other graphic arts fields are now on the market, and a little exploration of stores that sell supplies to draftsmen, architects, and printers will turn up several items of interest to photographers. For example, architects' plastic shading plates can be sandwiched with a negative or multiple printed to give a regular coarse pattern, and add dramatic emphasis to the right picture. These plates are thin plastic wafers a few inches square, and cost about a dollar each. Pattern No. 97, made by the Technygraph Company, was used to make the photogram design in Fig. 3-20. The Mackie line makes the pattern more intricate when it is solarized. Also for architects' use are dry transfer materials in the form of 10″ × 15″ transparent sheets bearing patterns, letters, numbers, trees, cars, and many other things that an architect might want to include in his plans. These designs are meant to be transferred to paper by rubbing on the back with a pencil or stylus, and can be used in this way only once. However, they serve very well as texture screens, and can be used photographically over and over again. They are quite inexpensive, and a good stock of designs can be had for only a few dollars. We have used three different brands, Lettraset, Instatype, and Prestype, with equal success, and show some of the results in Figs. 3-21, 3-22, and 3-23.

For less regular and formal patterns, the best approach is to use your camera. Whenever you see a brick wall, a weathered fence or an interesting pattern of glass or tile, take a picture or two, or more, at different magnifications, and add the negatives to your file. Slight underexposure is called for, because these are to be used in sandwiches with other negatives. Obviously, there is no sharp dividing line

3-19

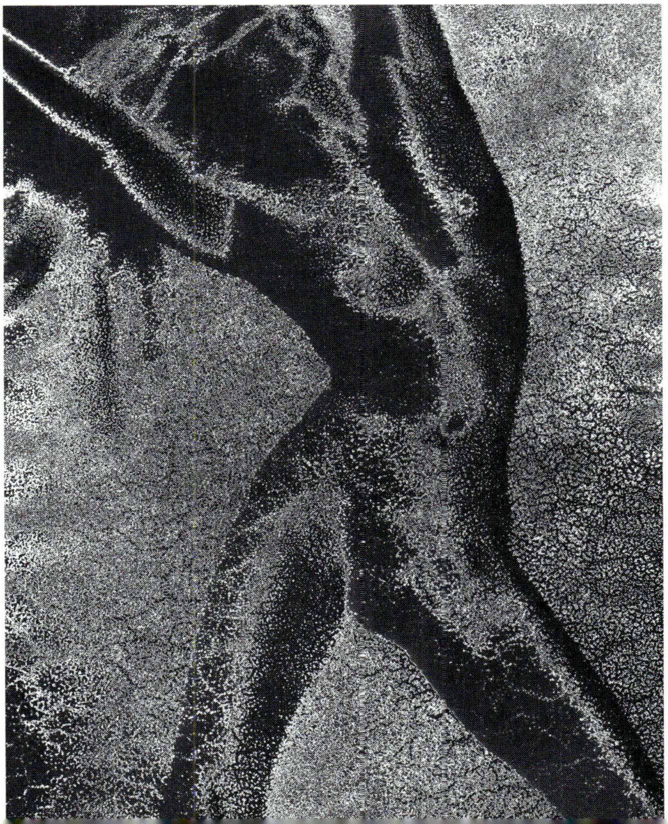

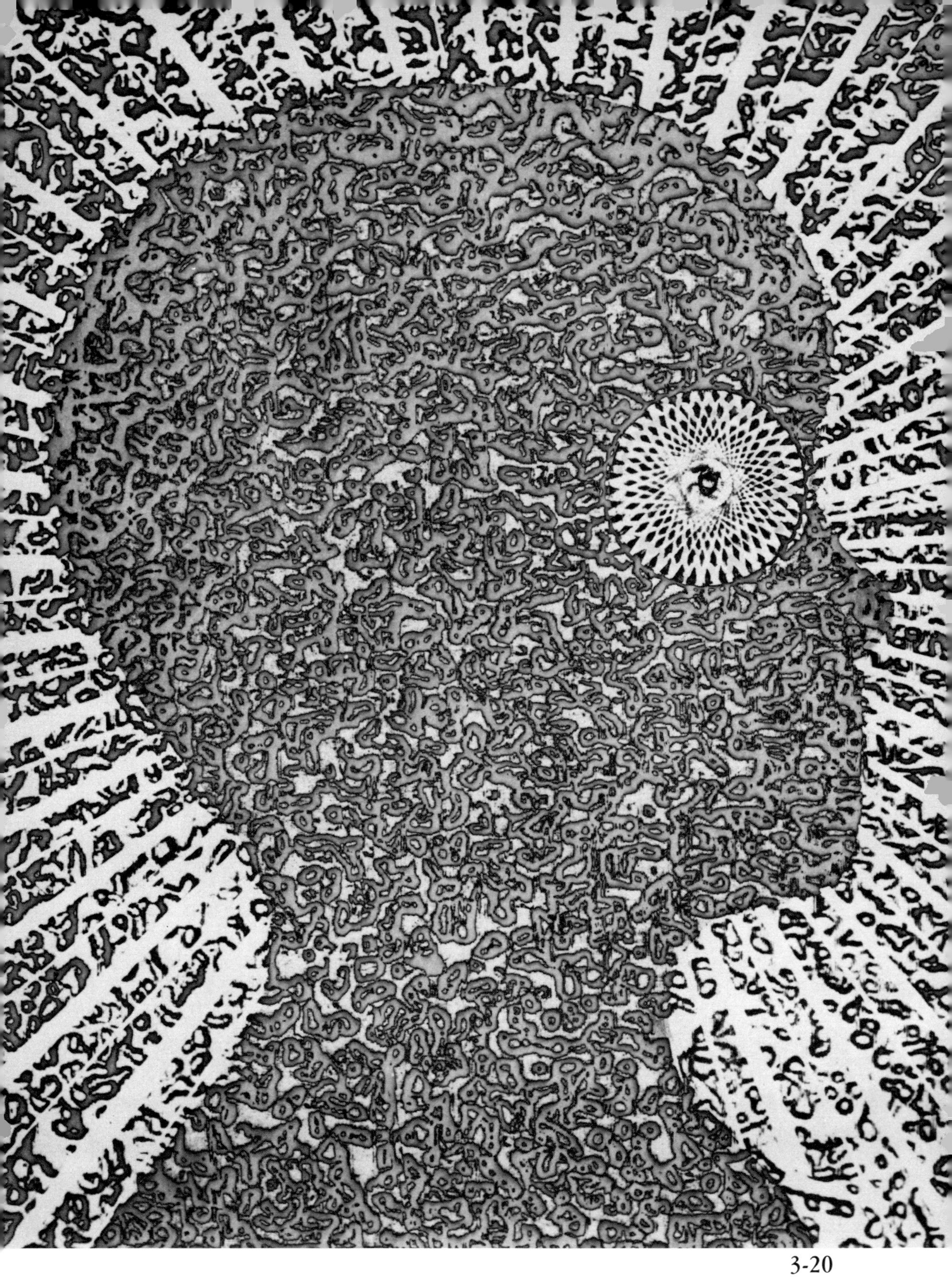

3-20

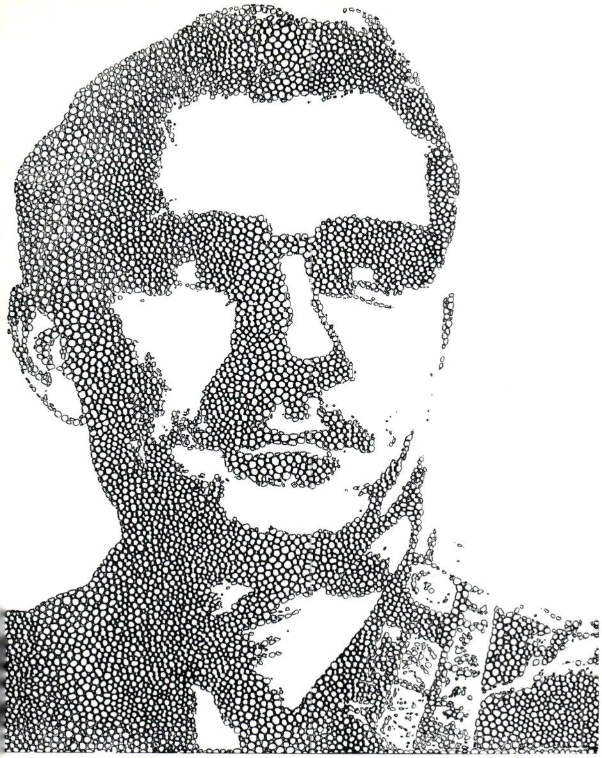

3-21

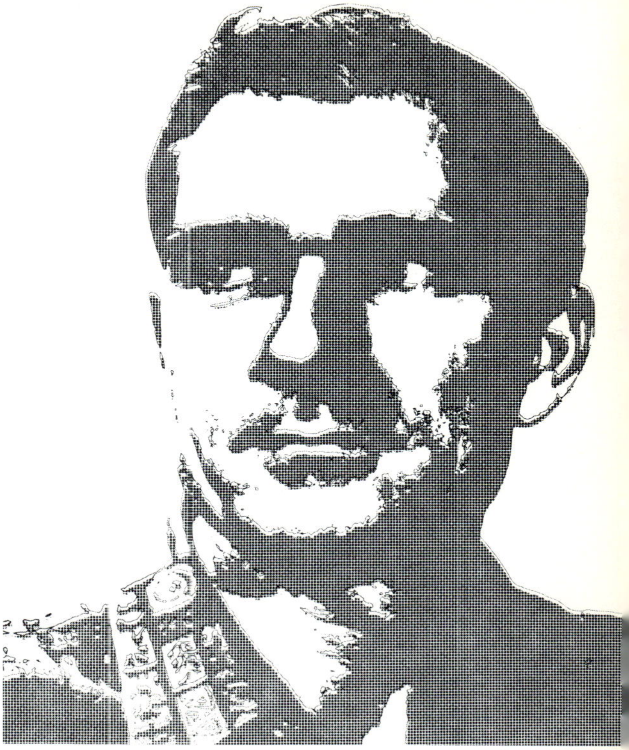

3-22

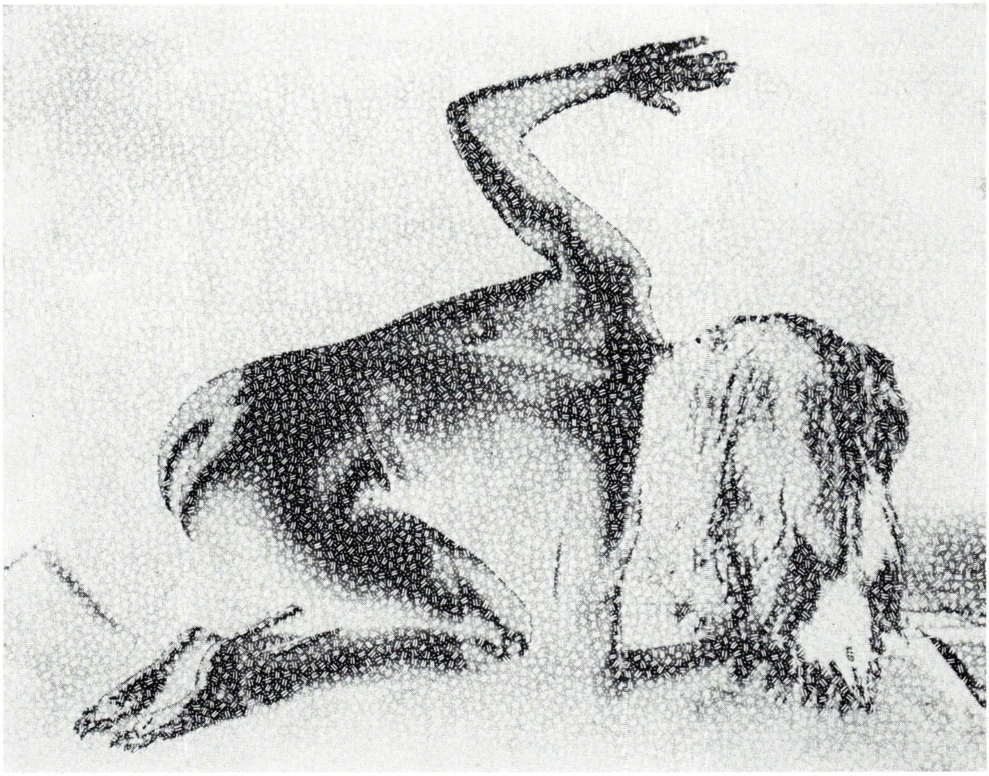

3-23

between a texture and a pattern but a general distinction would be that a texture is a fine grained effect that is unrelated to the design of the picture but influences its mood by a suggestion of surface character, while a pattern becomes an integral part of the composition.

Many natural and manufactured materials can be used directly in the enlarger, or on top of the print. Leaves, cloth, plastic, gauze, tissue, or glassine (Fig. 3-24)—almost any transparent or translucent material will work. Also, you can apply liquids, creams, powders, inks, or oils to glass sheets to create an endless variety of effects. Shaving cream, toothpaste, soap, syrup, paint, glue, sand, or mud are all good possibilities. India ink, for example, will dry with a lot of fine cracks when painted on glass or film. The effect will be exactly like the reticulation of the film emulsion that sometimes occurs due to a drastic change in the temperature of processing solutions.

Adding solarization to the process dramatizes the texture and opens up more design possibilities. It can be employed at any stage of the production with different effects on the results. The texture pattern or the print or both can be solarized separately and then combined, or they can be added together and solarized as any combination of positive and positive, negative and negative, or positive and negative. As you experiment, you will become more and more aware of the unending variety of visual effects under your control.

3-24

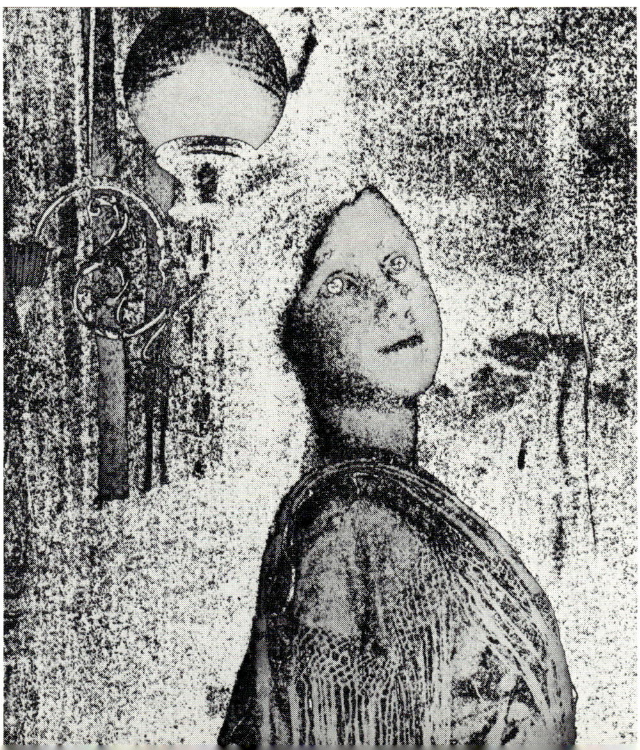

BROMOIL

As its name suggests, the bromoil is a combination of oil-base ink with a bromide print. One of the more elegant methods of print production in the early part of the century, the process has few practitioners today. A real bromoil is made by bleaching a bromide print in a tanning solution, which hardens the gelatin according to the image while removing the silver. When the bleached print is placed in water, the gelatin swells in the highlight areas, giving a relief image. Pigment is applied in the form of a thick greasy ink that is daubed on with a special brush while the print is wet. The ink will adhere to the hardened shadow areas but be repelled by the swollen highlight areas, which have taken up water. A great deal of control is possible during the inking stage, and an experienced worker can produce prints of great beauty. One characteristic of a bromoil is the delicate stipple of the brush marks, which gives open shadows and a soft tonal scale, more like a drawing than a photograph. Unwanted or conflicting details are left out, while the important ones are emphasized by careful shading. The color of the ink is a matter of choice, but a warm brown-black tone has been most traditional.

The brush stipple can be achieved in a pseudo-bromoil by printing through the back of the paper, by treating the paper negative with the abrasion tone method described in Chapter 2, or by a combination of both. Brief immersion in a brown toner will add a touch of authenticity, but the toning should not be overdone.

The original paper negative can be on any surface, but should not be of very high contrast (although too high contrast can be reduced by fogging). The pseudo-bromoil should be printed on a rough-textured paper such as charcoal, portrait proof, or tapestry, through the back of the print. A black printing surface should be used beneath the paper in order to minimize back reflection, since the emulsion of the printing paper is downward.

Solarization is not necessary to produce a good imitation bromoil, but in some cases it can be used to great advantage. In our example (Figs. 3-25A to E) it was used to dramatize the clouds and to subdue the unwanted detail of the telegraph poles. There were three steps: First, a straight print was made on No. 3 paper; second, the straight print was reprinted as a negative onto No. 6 paper and lightly solarized; finally, a print was made through the back of normal-grade paper with a tapestry texture. The straight print is shown in Fig. 3-25A, the solarized negative in Fig. 3-25B, and the final print in Fig. 3-25C. If the negative had not been solarized, the result would have looked like Fig. 3-25D. While making this fake bromoil, we were struck by the winter-like quality of the solarized paper negative, so we made a lighter version of it to emphasize the appearance of snow (Fig. 3-25E).

Most "bromoils" are not the original prints, but are bromoil transfers, which are ink impressions made from the original onto non-photographic paper. The bromoil technique was most used by the pictorialists, and their favorite subject matter tended to be of a romantic nature. Pastoral scenes, old barns and churches, fishing boats, and village scenes are typical bromoil subjects, and the soft, diffuse image serves them well. We used a shot of a beautiful eucalyptus tree we had found growing beside a country road in Queensland, Australia, for our bromoil imitation.

3-25A

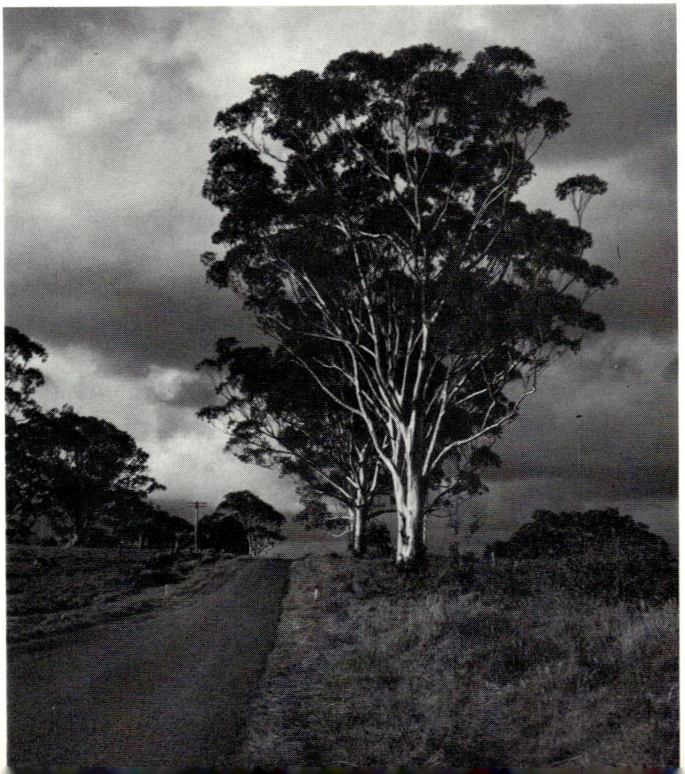

3-25B

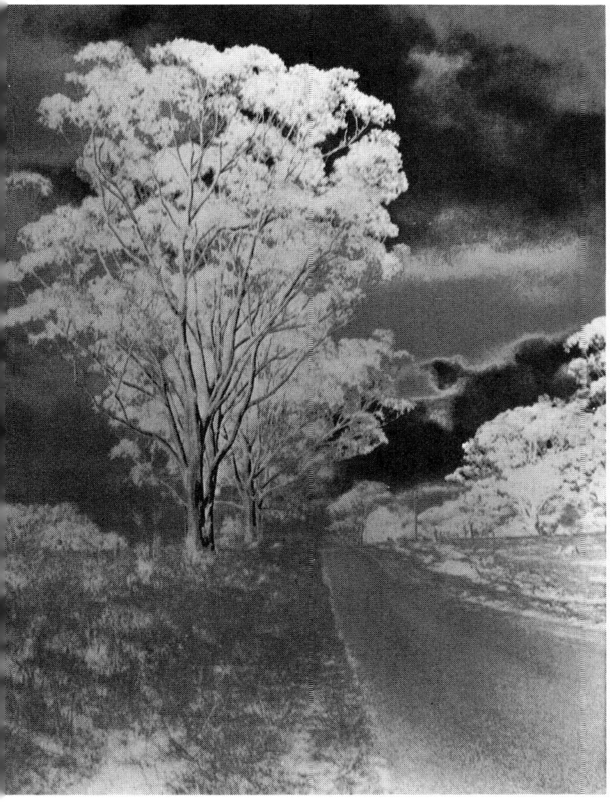

3-25C

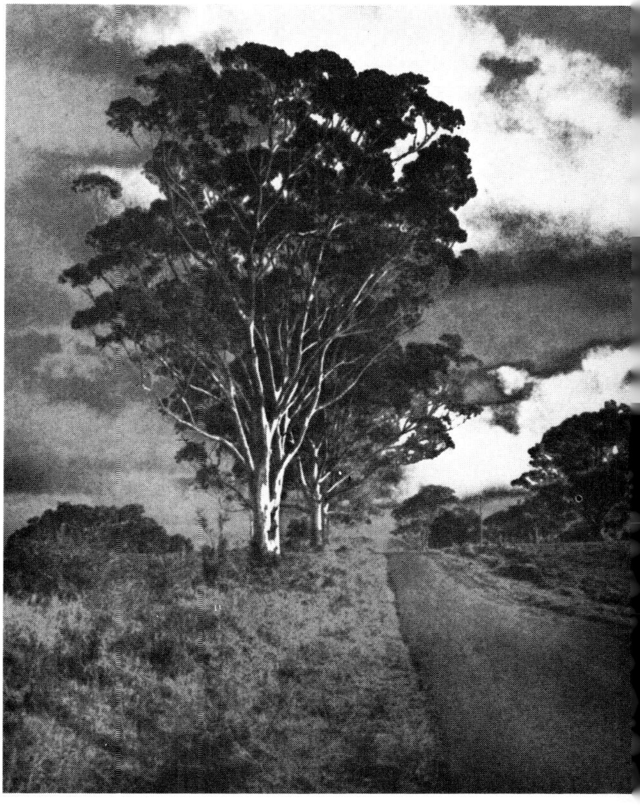

3-25D

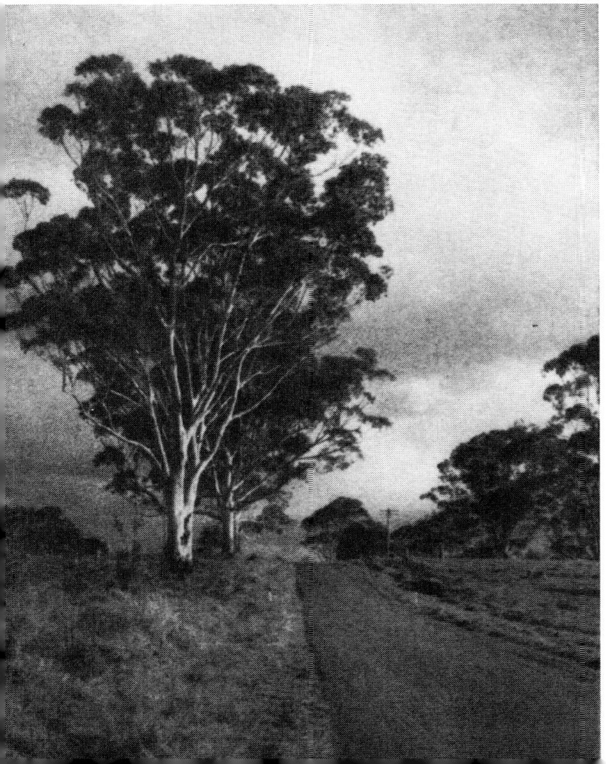

3-25E

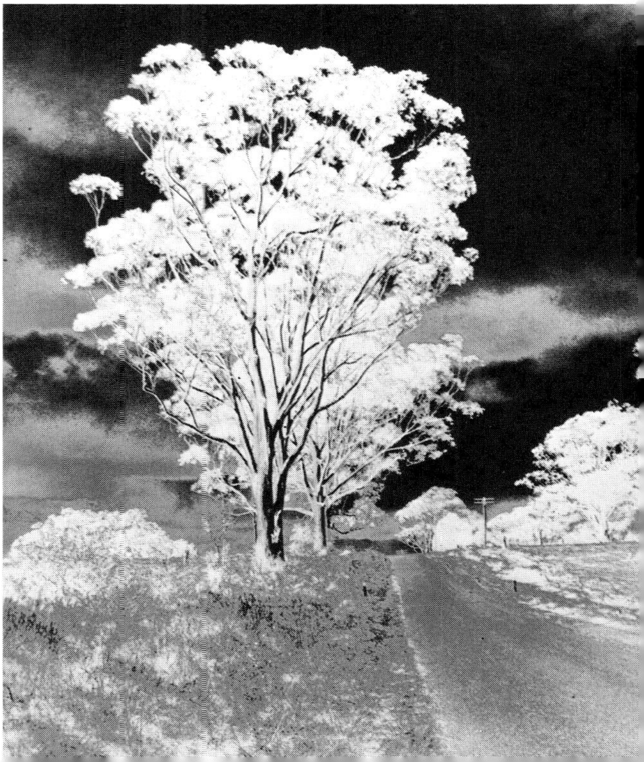

TURNING DAY INTO NIGHT

Reversal by solarization can be selectively applied to different parts of a picture if they are very different in tone. This allows us to print a picture taken in daylight in such a way that it seems to have been taken at night. This is, in fact, the most usual result of simple solarization. A bright, daylight sky turns dark or black during solarization. In the usual case, the tones of the subject also reverse to some extent, and the result takes on an unreal appearance, rather than that of a night picture. However, when the exposure and development are such that the subject is almost fully developed before the sky begins to darken, the proper solarizing exposure can darken the sky without appreciably affecting the subject. Also, there is often a halo effect around the subject, which makes it stand out against the background. The photograph for Fig. 3-26 was taken in the early afternoon against a hazy bright sky with the girl and rocks almost in silhouette. The three prints show three degrees of solarization. In Fig. 3-26A the first development was full, and the second exposure was just enough to lower the contrast of the No. 6 paper so that a normal print could be made. In Fig. 3-26B the first exposure was slightly shorter and the second exposure was slightly longer. The most interesting feature of this example, aside from the increased clarity of the figure detail, is the black line delineating the girl's body, making her appear to be a cut-out image pasted on the picture. This effect is somewhat unusual, and will be discussed further in Chapter 5. The final picture (Fig. 3-26C) carries the transformation all the way. Slight reversal has occurred on the model's right foot, but otherwise the tones are natural, with highlights and shadows in their proper places. The rocks are illuminated and the sky is dark, with a slightly luminous quality around the figure, which has a fine white outline.

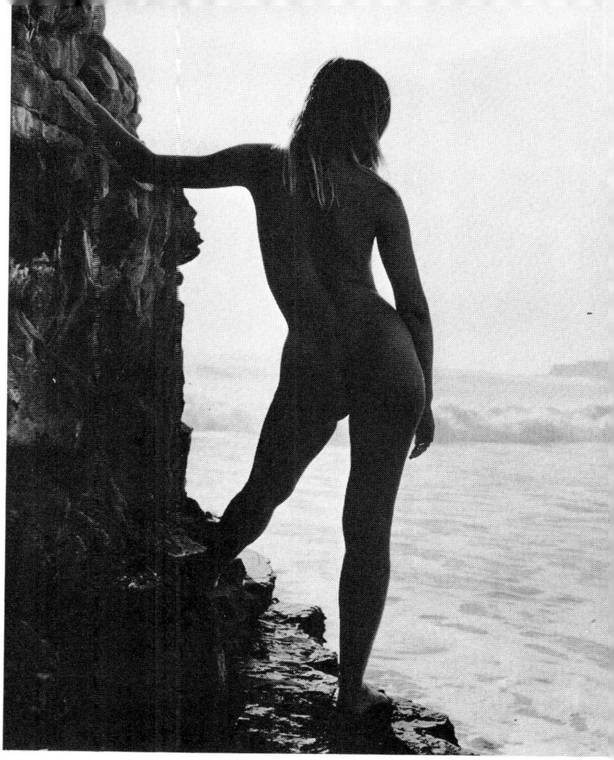

3-26A

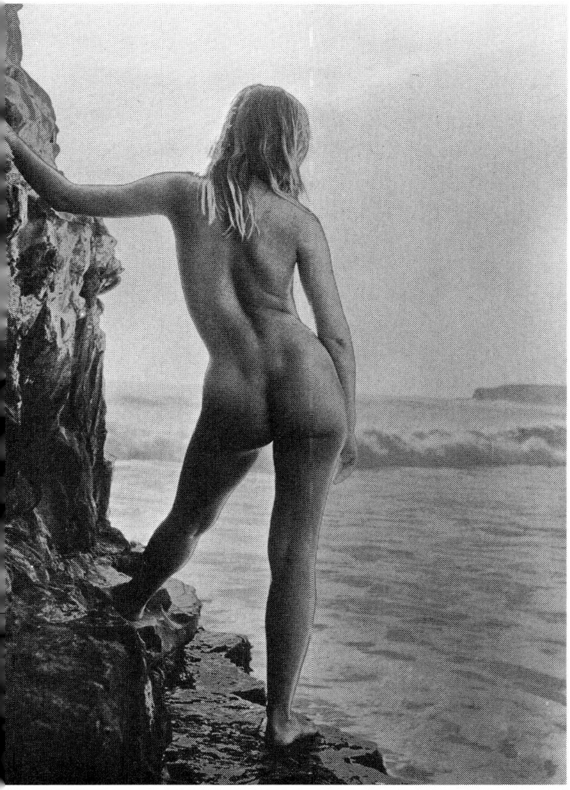

3-26B

3-26C

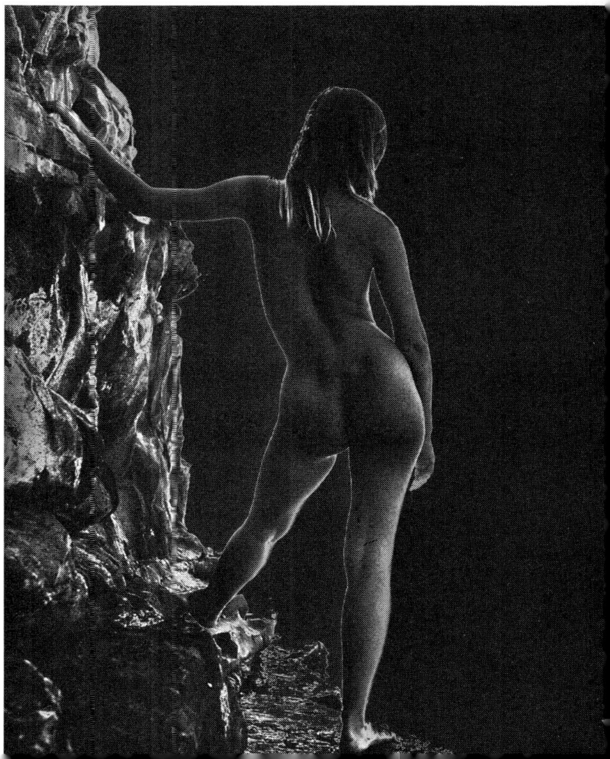

TURNING NIGHT INTO DAY

We have just seen how easy it is to turn day into night. At once, the question comes up: Can we also turn night into day? The answer is, "Yes," but the process is not quite so easy. There are two ways to achieve the night-into-day effect, but neither of them results in a full-scale print, and they both require careful control to achieve a believable result. The negative of a picture taken at night will have a clear background with normal density in the foreground subject. The problem is the same as before: Reverse the background without reversing the subject. The difficulty stems from the fact that, in this case, the background will develop out first. We can proceed as before: Let the background develop almost fully, then solarize to reverse the subject. This will give a negative image of the subject on a black background. When printed, it will give a positive image of the subject on a white background—that is, a daytime picture. That sounds simple enough, but there is a catch. The details of the subject may have some densities comparable to the background, and so the subject will not reverse uniformly. Also, the final print is quite harsh—contrasty and lacking in tonal range. We prefer a different procedure, one which depends upon a peculiarity of the solarizing process. Consider these three cases: (1) A print that has been normally exposed and fully developed will not reverse when solarized, but will only develop the edge effects and silver highlights; (2) A print that has been slightly underexposed and only partly developed will reverse and become a negative when solarized; (3) A print that has been drastically underexposed, partially developed, and given a brief re-exposure before part of the image has yet become visible will develop with the previously invisible image as a positive of low contrast.

The eucalyptus tree, Fig. 3-27A, was photographed at night by electronic flash, and has been given the latter treatment. The original print was very much underexposed so that after a full minute of development the background was just becoming visible, and no details of the tree had yet appeared at all. A ten-second exposure to a 25-watt lamp was given after about one and one-half minutes in the developer.

After another half-minute, the tree details began to appear, developing much faster than the background, which remained a light grey. Full development gave a positive image of the tree against a lighter sky—a low contrast picture, but no longer a night shot. The contrast was improved by reprinting twice to the second generation image that is shown in Fig. 3-27B. We see that our final image is now a daytime picture, but we can hardly claim that it has the photographic quality of the original. Rather, it has taken on more the quality of a charcoal drawing, an effect you may or may not find pleasing.

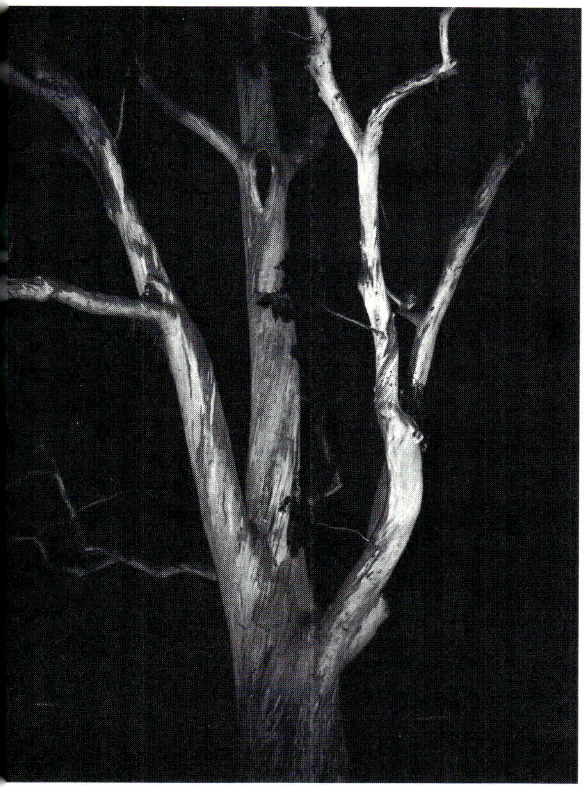

3-27A

3-27B

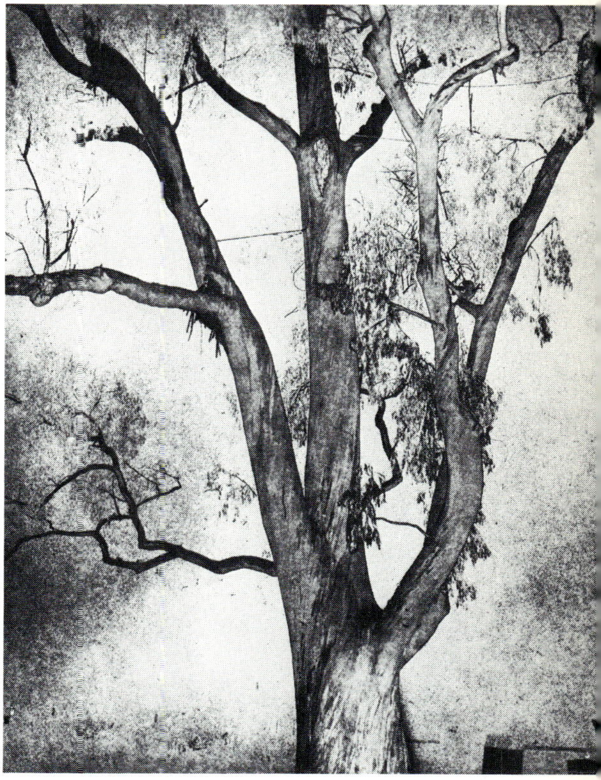

BAS RELIEF

In conventional photography, fake bas-relief images are produced by sandwiching a positive transparency with the negative of the same image slightly out of register. This produces a print showing the image with a white (illuminated) outline on one side and dark (shadow) outline on the other, giving the appearance of a raised surface characteristic of bas-relief sculpture. A similar effect can be achieved with paper negatives, when positive or negative or both have been solarized. Because of the mixed character of the solarized image, being neither all positive nor all negative, the final image is likely to have some unique features not found in the conventional bas-relief result. For instance, the shadow line is likely to appear on both sides in some details and not at all on others. The general effect is often a strange distortion of the direction and quality of the lighting.

Registration is the most difficult problem when superimposing images in photography, and it is much easier to make them slightly out of register than to get them exactly in register. When dealing with paper negatives, certain precautions should be taken to insure that the images remain the same size. An 8″ x 10″ print can change size by as much as 1/16″ during the processing operation, and the exact amount depends upon the kind of paper used as well as the temperature, acidity, and duration of the various liquid baths. To avoid trouble from this source, it is wise to use the same kind of paper for both positive and negative, and treat the two exactly the same while processing. For effective bas reliefs, the two images should be quite different in character, one being more negative while the other is more positive. A solarized print with a straight print is a good combination. They should both be printed face down, and the result will be slightly different depending upon which one is on top. Place them face down on the light box, and arrange them until you get the effect you want. Tape them together at one edge so that they will stay in place, while you print them in the usual way on the printing frame. Their printing time will, of course, be somewhat longer than usual, and it is essential that they be held down tightly to avoid "soft" spots in the print.

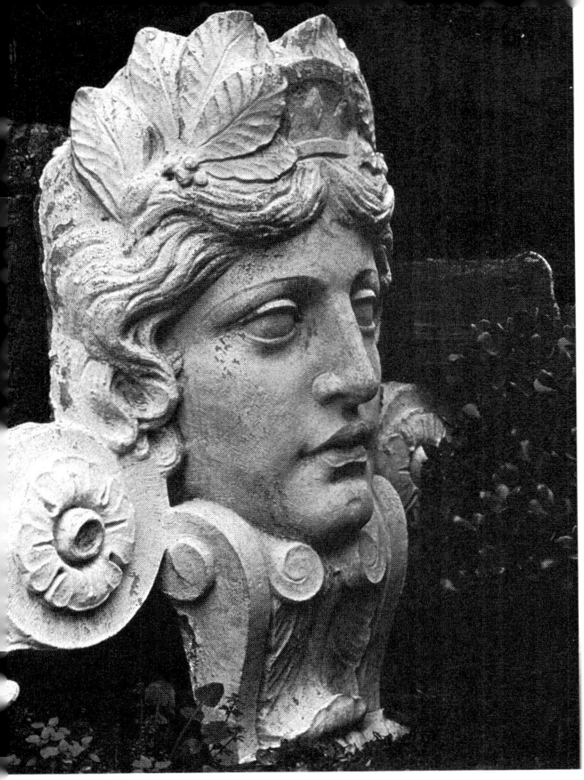

3-28A

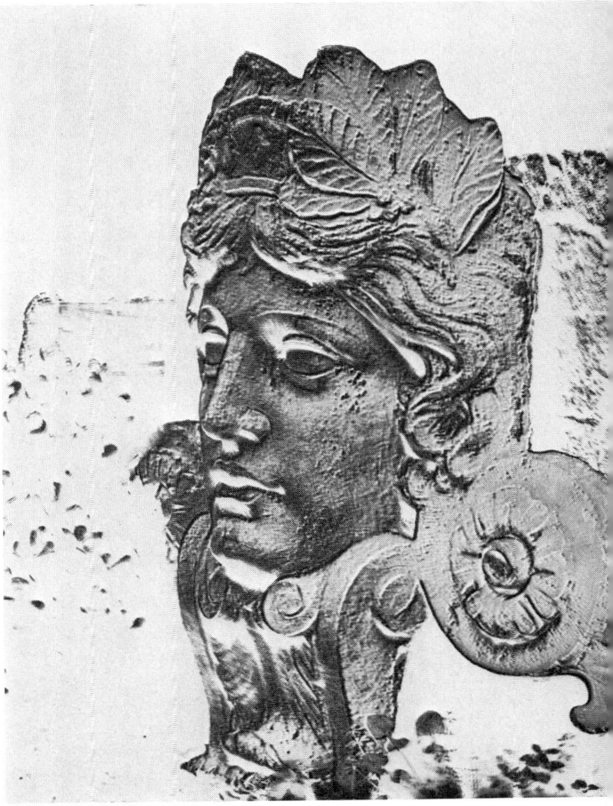

3-28C

3-28B

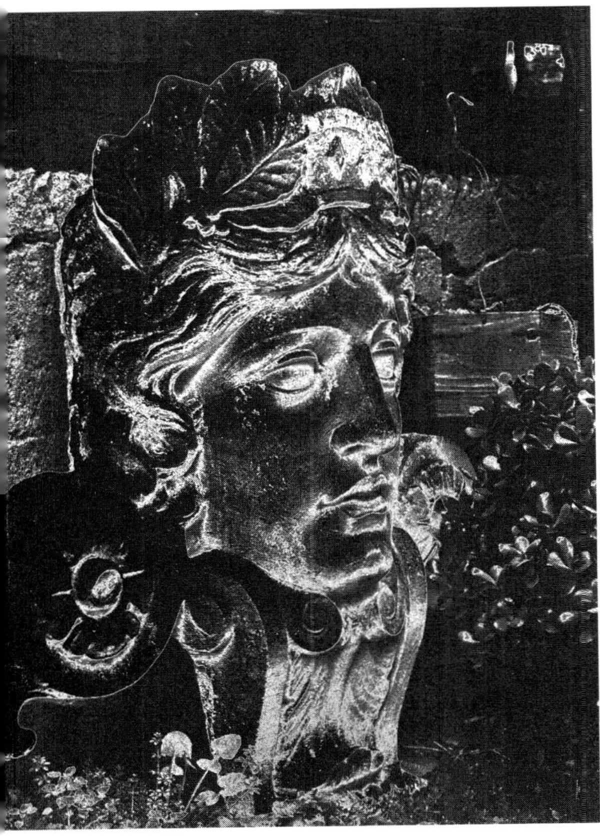

3-28D

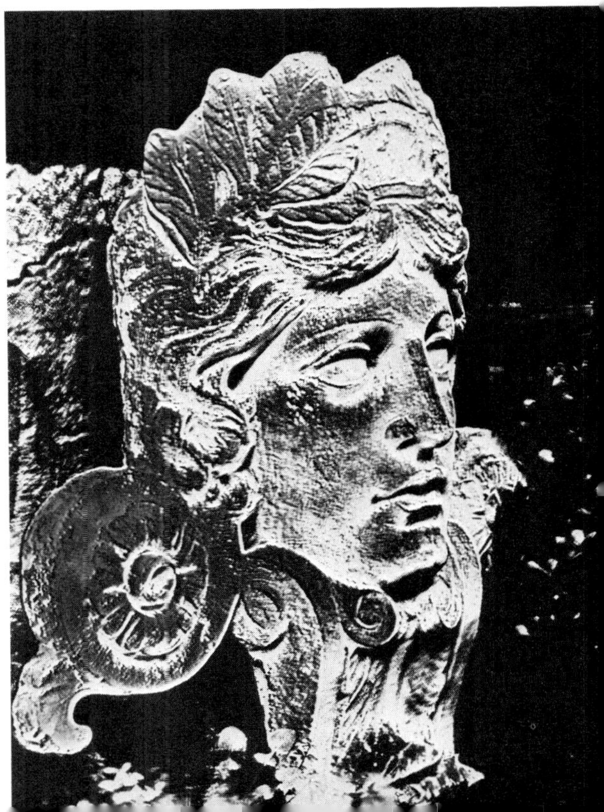

The sculptured head was photographed on a very overcast day, but the flat negative made a good print on No. 6 paper (Fig. 3-28A). An identical print was solarized (Fig. 3-28B) and the two were printed together to give a strongly negative bas relief (Fig. 3-28C). The more positive version in Fig. 3-28D is just a print from Fig. 3-28C. The English cottage in Fig. 3-29 has been given the same treatment, but both prints were negative, one of them solarized, so that the result is a good positive. The registration was almost exact in this case, so the relief effect is evident only in the fine details.

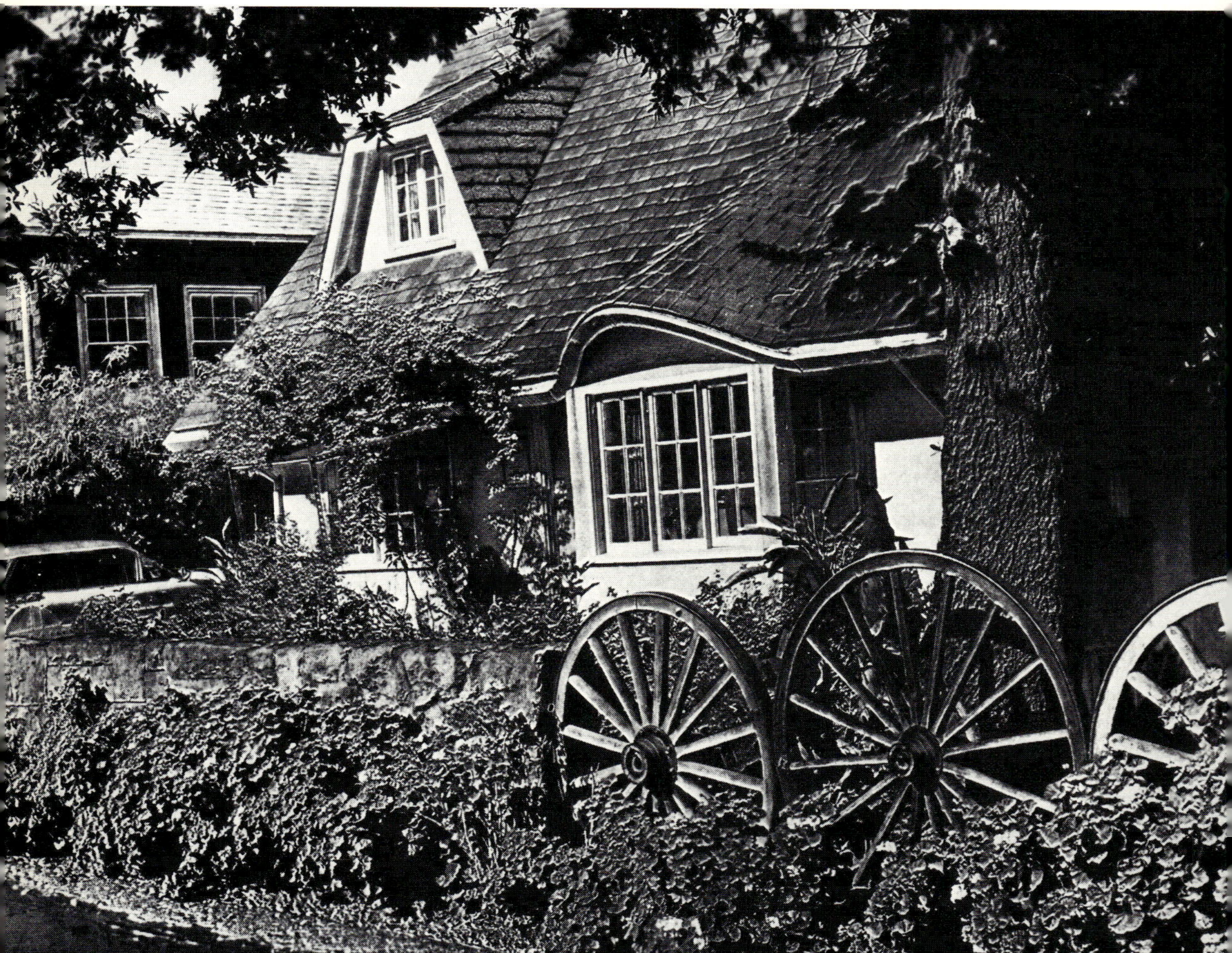

3-29

TONE SEPARATION

A normal, full-scale photograph ideally contains a range of tones, varying in imperceptible steps, from pure white to pure black. For impact and simplicity, however, it is sometimes desirable to reduce the number of intermediate tones to one or two distinguishable steps between black and white. The process of doing this is called posterization, or tone separation. To accomplish this photographically requires dropping out all the intermediate tones, and then replacing the ones that you want. This calls for some kind of double printing and for exact registration. The double-printing requirement can be met by solarization, and exact registration is merely a matter of accurate guide marks or pins on the enlarging easel.

Begin by making a high contrast print and solarizing it so that it is largely negative with strong Mackie lines. (Remember, if you want the print to reverse, you must give the second exposure rather early in the development phase.) After the solarized print is dry, take a print of it to use as a registration guide on the easel. Turn the negative over in the enlarger and register the image with this second print on the easel. When the alignment is correct, put aside the second print and take up the solarized print. Place it emulsion-to-emulsion with an unexposed sheet of paper onto the printing frame and expose, but do not develop. Instead, place the undeveloped sheet under the enlarger in correct orientation with the negative and make an exposure just sufficient to bring up a grey tone when developed. From the solarized print, you have produced only black and white, and from the enlarger exposure you have introduced the intermediate tones. The balance between these exposures determines the final distribution of tones, and some experimentation is usually required to get the result you want.

The examples show the steps in this process, beginning with a soft-focus portrait (Fig. 3-30A) that has also been blurred by camera movement. Note that the final image (Fig. 3-30D) is a positive even though it has a reversed Mackie line. The two buildings that have been given similar treatment are of a decidedly different character. The Eighth Street meat market (Fig. 3-31) is in downtown Oakland, California, while the other building (Fig. 3-32) is part of the University of Sonora, Mexico.

3-30A. The high contrast, soft-focus portrait.

3-30B. The solarized print, largely negative with strong Mackie lines.

3-30C. The print made from the solarized print, prior to exposure in the enlarger.

3-30D. The final image after the enlarger exposure.

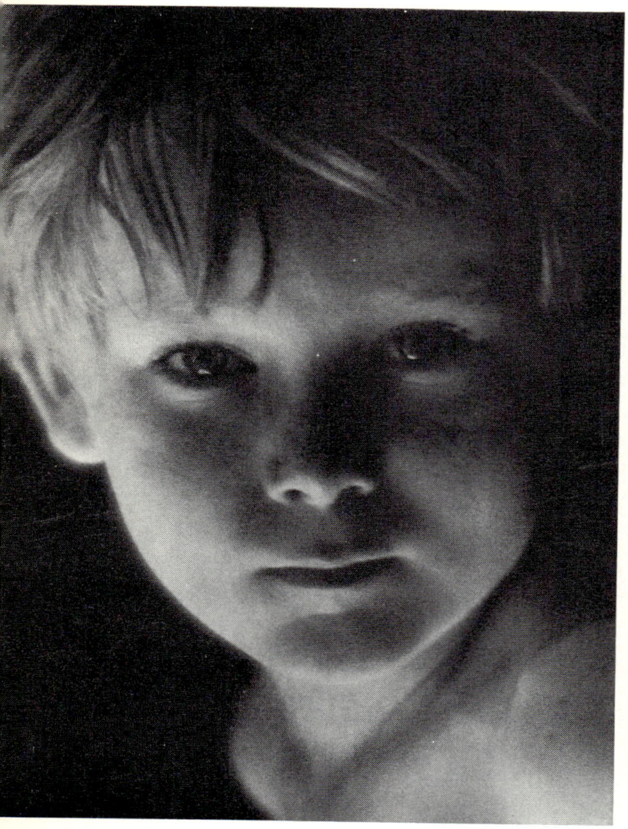

3-30A

3-30B

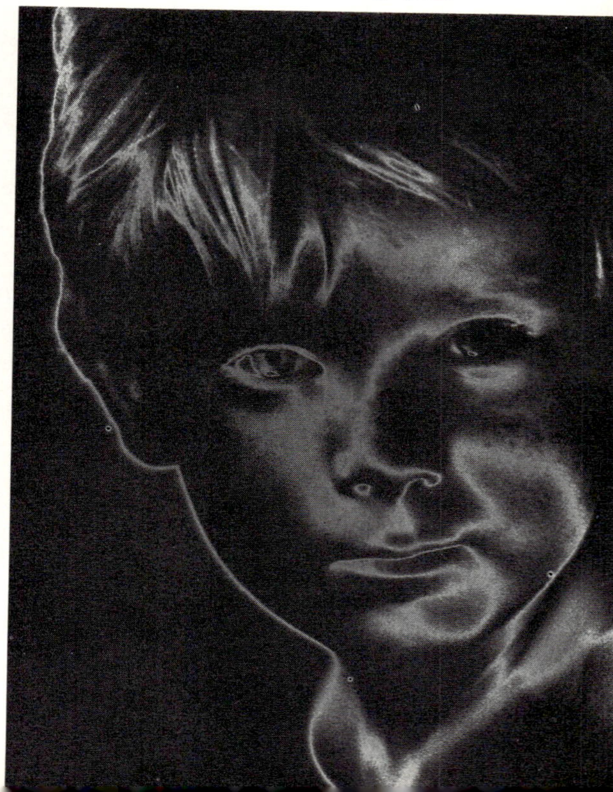

3-31

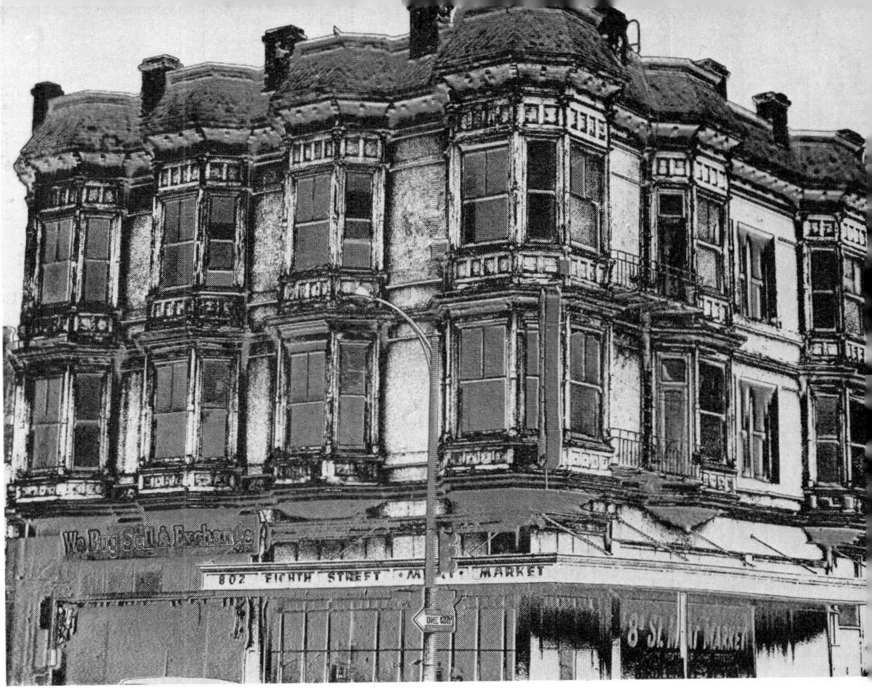

3-30D

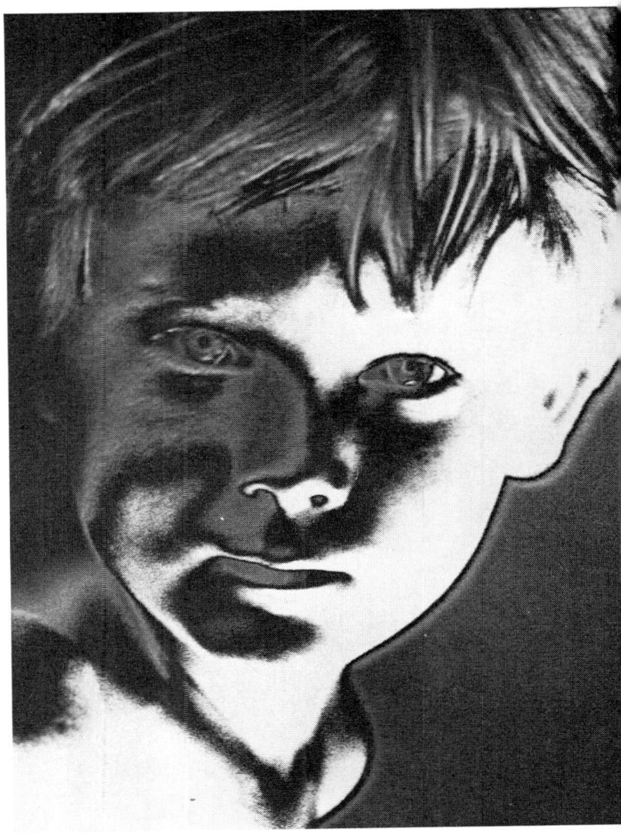

3-30C

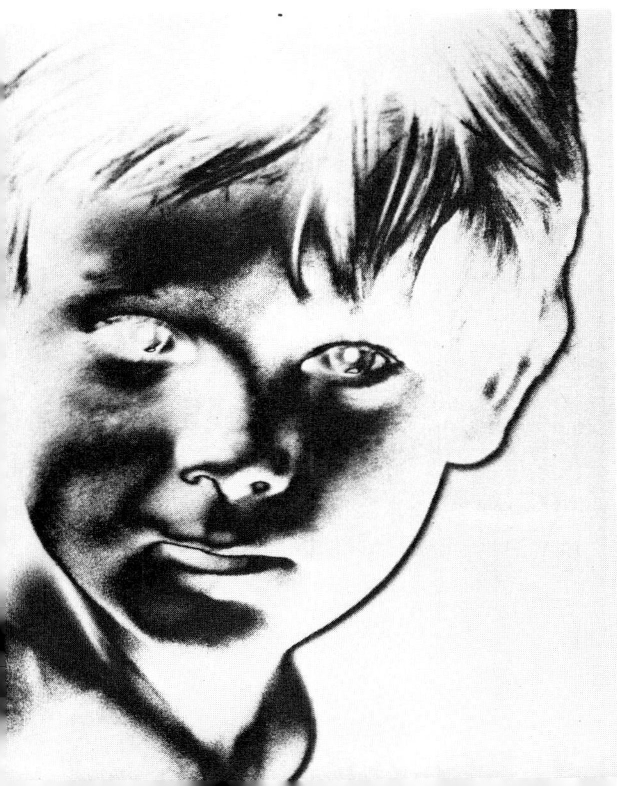

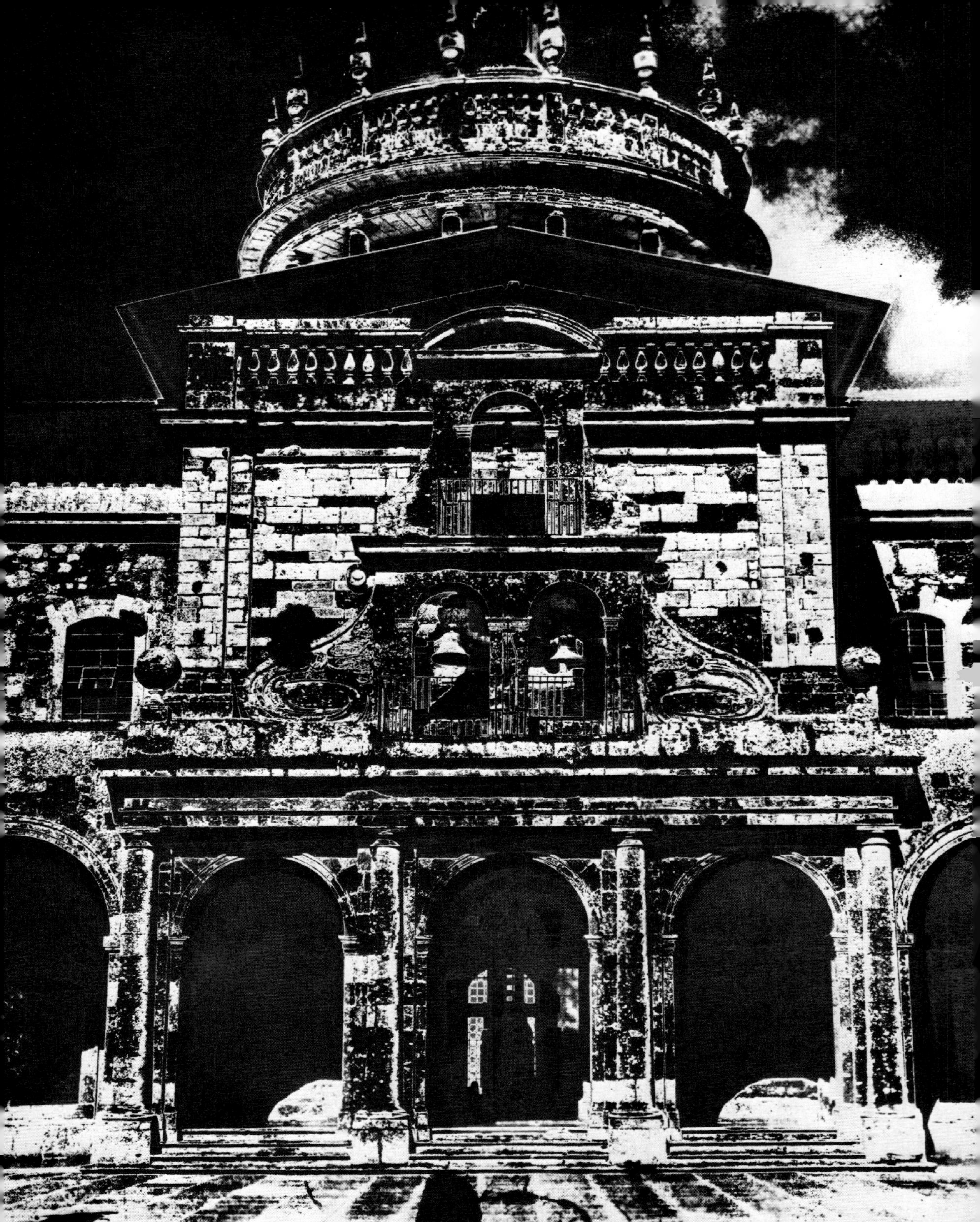

3-32

GENTLE SOLARIZATION

In most of our solarized pictures the subject matter has been severely distorted in some way, and is often almost unrecognizable. The aim has been to create a different picture, not just to embellish the existing one. We have found a technique, however, that is not nearly so drastic, and the result of which is simply the improvement of picture quality. We have called the technique "gentle solarization," although it could just as well be called "the easy way," because it is no trouble at all, and is a real pleasure to do. The procedure is this: Exchange the overhead solarizing light bulb for one of approximately 15 watts. Make a print on No. 6 paper, develop it fully, and then, while it is still in the developer, turn on the light and leave it on! Watch the print carefully, and when it reaches the desired quality, transfer it to the stop bath and then to the fixer. Don't turn off the light and don't worry about fogging the print. Remember, you are working with "old brown" developer, which takes about three minutes to produce a good print. It works slowly enough for you to stop the solarization at any point, and, with the weak white light, the changes in the print take place over a time span of several minutes. Instead of fogging the print, this process actually enhances the contrast of the fine details, increases the apparent brilliance, and improves the pictorial quality of the print. Compare the three straight prints in Figs. 3-33A, 3-34A, and 3-35A with their counterparts in Figs. 3-33B, 3-34B, and 3-35B, which have been subjected to the gentle solarization treatment. The pleasure of this procedure is in watching the progress of the print with the light on. We find it so enjoyable that we do many of our prints this way, regardless of the degree to which we want them solarized, changing only the light bulb to suit the speed and the desired result. In general, a stronger light bulb works faster and produces a deeper black, but the quality of the result also depends very much on the time of the first development.

3-33A

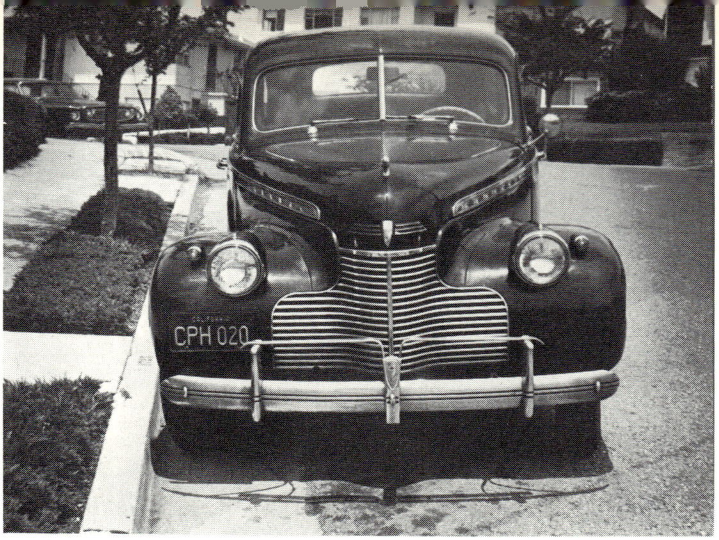

3-34A

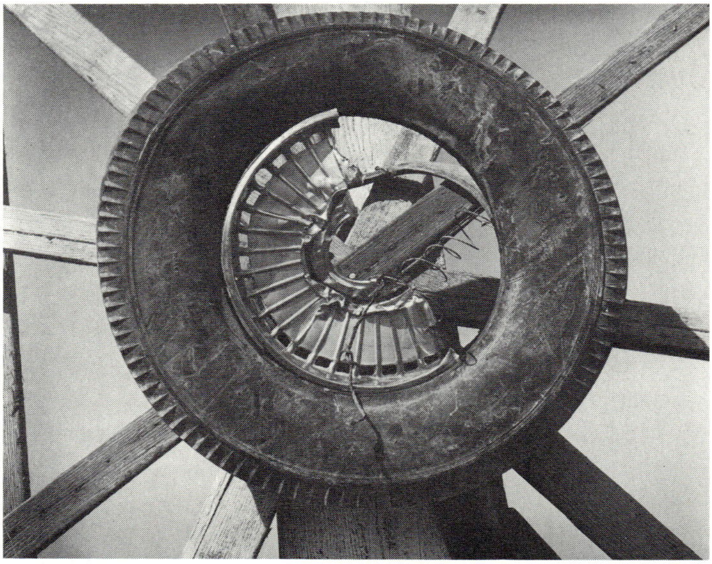

3-35A

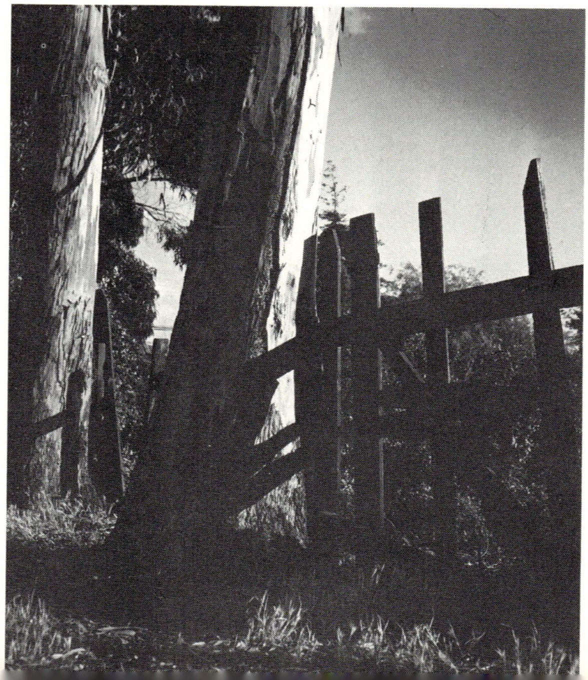

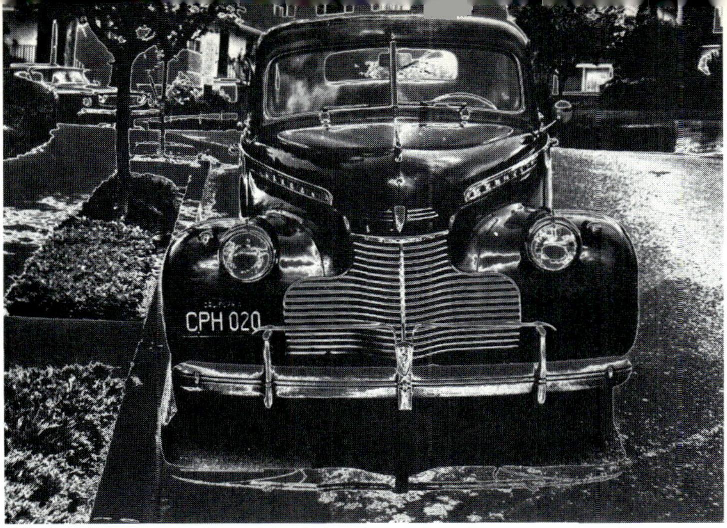

3-33B

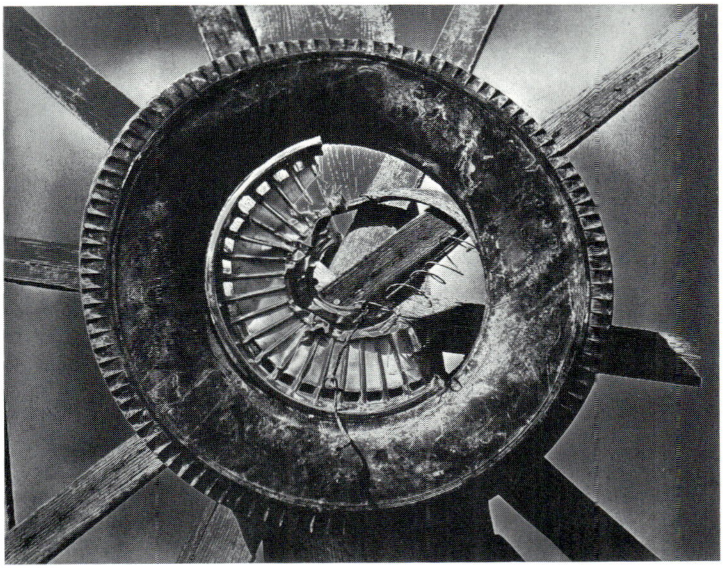

3-34B

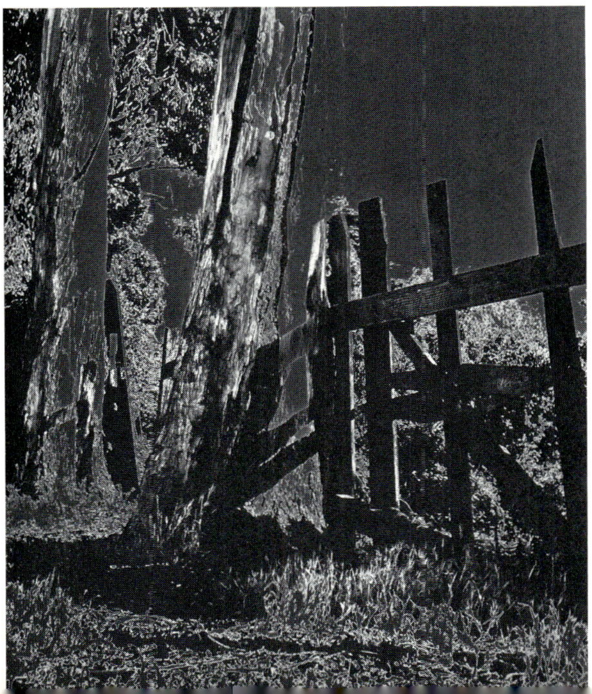

3-35B

BLEACHING

All well equipped darkrooms should stock a bleach, not just for correcting mistakes but to create and improve photographic images. The ability to properly use a bleach is one of the necessities of the photographic craft. Detailed formulae for bleaches for different purposes and careful directions for their use are found in several good books on printing and enlarging, where they are usually listed under "reducing" rather than "bleaching." As usual, we limit our remarks to our own experience and the applications to solarized prints, referring you to other works for the more general case.

We have already mentioned the use of a household bleach in Chapter 2 for working on paper negatives. That was a strong, non-reversible bleach used on a dry print, and the print was not intended to be the final image, but was only to be used as a negative. We are now talking about reducing the density of a good print, either as a whole or in part, the aim being to improve the quality of the image. For this we need a controllable and reversible bleach that will not stain nor damage the print in any way. As often happens, a simple formula has turned out to be best:

Potassium Ferricyanide	20 g
Potassium Bromide	20 g
Water to make	1000cc

These proportions are not critical, and the 20-20 ratio may be varied considerably without serious effect. The dilution determines the speed of bleaching, and must be decided according to the particular use intended. Always test the strength on a junk print before risking a good one.

The most common use for a bleach bath is the general reduction of density in a dark print and the clearing of veiled highlights. Since solarized prints often suffer from both of these faults, bleaching is a made-to-order remedy. For simple reduction of density with slight change of contrast, a proportional reducer is required. This means that the bleach removes silver in proportion to the amount present in the image, thus reducing less of the highlights and more of the shadows. To increase the contrast, a cutting reducer is required, since this removes the same amount of silver from all parts of the image; the highlights become clear while the shadows still appear dark. Remember, we are talking about prints here, not negatives! Whether the bleach is a cutting or a proportional one depends somewhat upon its dilution, a strong solution being more of a cutting type.

In practically all cases of interest to us here, the aim is to increase contrast while reducing the overall density of the print, so we tend towards a relatively concentrated bleach bath. Using the above formula as the stock solution, a reasonable bath is made by adding four parts of water to one of stock. The print to be bleached should be well fixed and washed. If it has also been dried, it should be soaked in water for at least 30 minutes before placing it in the bleach bath. Agitate the print in the bath, watching it carefully, particularly the highlights. Just before it has reached the ideal condition, remove it to a water tray for a quick rinse, then place it in the fixer for a few minutes. The final wash should be for one hour in running water. Examine the print in the first rinse water before placing it in the fix. If it is too light, and you decide that you have carried the bleaching too far, all is not lost. Just rinse the print well, and put it back in the developer until it has come back to where you want it. Then fix in the usual manner. If you decide that the bleaching is insufficient, the print can be returned to the bleach bath for an additional treatment. Some clearing takes place in the fixing bath, so it will take a little experience to be able to judge exactly when the bleaching has reached the proper stage. This is why it is wise to use a test print first.

If the process goes too slowly, increase the concentration of the bleach. If it goes too fast, reduce the concentration. The proper speed of bleaching is a matter of your temperament and patience. As you gain experience, you will probably want to use a stronger bath and get the work done faster. Fig. 3-36 shows the effect of the bleach on the left half of a print.

If the reduction is carried too far the print quality will be degraded, and very severe bleaching may produce discoloration. Occasionally, a change in color can be deliberately produced and used to good effect. The silver deposit due to the second exposure is of finer grain than the image produced by the first exposure. This makes the solarized part of the picture relatively more susceptible to bleaching, and

3-36

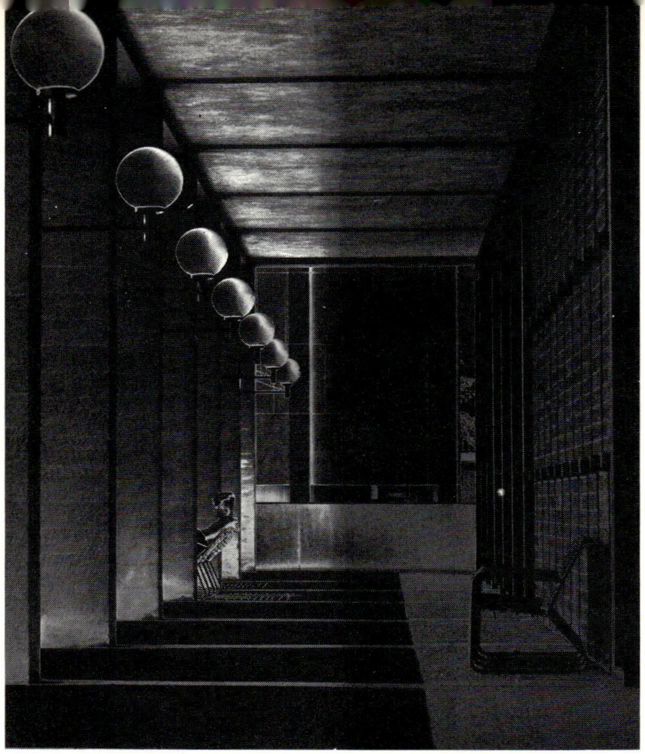

at the same time imparts a warm brown tone to the bleached area. Thus, a solarized print that has been strongly bleached often exhibits a brown tone in the solarized area while the original image remains black. This effect is favored by a relatively light first exposure with full development followed by a strong second exposure and short development.

Most instructions for the use of a ferricyanide bleach caution you to use a solution of plain hypo, sodium thiosulfate, instead of a regular fixer. Often no reasons are given, but sometimes there is a warning about staining if you use fixer. We regularly use Kodafix for all our work, and have had no trouble with staining, but certainly, a plain hypo bath should work just as well.

One of the nicest things about this bleaching procedure is the fact that it can be reversed by redevelopment, provided the print has not been fixed after bleaching. Once the print has been bleached and fixed, it can not be redeveloped, but it can be returned to the bleach bath for additional reduction. A bleached print that has not been fixed can be redeveloped in any normal developer as well as any toning developer that you choose, and this turns out to be one of the easiest and best ways to tone prints. Some of our best results have been obtained by redeveloping a bleached print in

GAF Direct Sepia Toner, although it is intended to be a direct toner that does not require bleaching. According to the manufacturer's directions, it is not recommended for cold-toned papers, but we find it works beautifully with these papers when used after bleaching. It was used for all the sepia-toned prints in this book.

Another important use for a bleach is to remove unwanted parts of an image or to clear up the fogged borders that often appear when a paper negative has been printed too heavily. For this application, it is best to use the mixture full strength and apply it with a brush, cotton swab, or wad of cotton. Arrange one glass beaker with the bleach mix in it and another beaker containing fixer at normal strength. Have a bath of clear water handy, and a few swabs. Soak the print in water for at least one hour before working on it to ensure that the emulsion is uniformly swollen. Lay the wet print out on a sheet of plastic or the back of a developing tray and wipe off the surface water. For delicate work, apply the bleach with a fine brush; otherwise use a swab. Do not flood the area with the solution for it will diffuse into the image, but use only as much as you can control. Watch the action of the bleach, and when the desired density is reached, wipe the area with the swab dipped in fixer. Since the action of the concentrated bleach is rather rapid, it is best to work with a swab in each hand, one for bleach and one for fixer. Wash the entire print when the work is finished. For removing small black spots, it is not necessary to pre-wet the print. Just use a fine brush on the dry print, wipe with fixer, and, when all the black spots are gone, wash the whole print for a time depending on how saturated it is with fixer.

Bleaching is very worthwhile and not difficult to learn. However, it is recommended that you practice on some junk prints until you can recognize the abilities and limitations of the technique.

ADDING COLOR TO THE IMAGE

It is beyond the purpose of this book to mention all the possible ways in which color hues can be added to a photographic image, but our experiments with different materials and solutions in the study of solarization have led us to some observations on possible color techniques, other than conventional toning. One of these is simply staining by improper fixing. A print that has been dipped in the fixer to stop development is placed under a strong light or even out in the sunshine. It will turn a few hideous shades of brown, yellow, and purple—you have probably noticed the effect on discarded prints in your garbage can. This has been seriously suggested as a "new" variation on conventional methods, and a few examples of such accidental coloring have actually been hung in exhibits, but we prefer to forget this method until someone can show us that they can control the stains and produce more attractive colors.

A more honest attempt at color innovation is represented by the tinted bromide papers that have appeared on the market under such names as Color-Brome, Luminos Pastel, and so on. The paper stock itself is colored red, blue, green, yellow, gold, or silver, and so on, while the image that develops in the course of ordinary processing is the usual black, silver image. While processing, the paper appears tinted with the appropriate hue, but has a grey cast and looks rather dull. When placed in the fixing bath, however, the grey veiling fades away and the print takes on a very respectable color with adequate saturation. Our question about these colored papers was naturally, "Will they solarize?" The answer is, "Yes, and quite well," in spite of the fact that they seem to be available only in "normal" contrast grade No. 2. One might fear that the solarizing exposure would fog the tinted paper in such a way that the color hue would be seriously degraded and become an unpleasant muddy tone. This does not seem to be the case. The hue remains quite strong and darkens only a reasonable amount without unpleasant veiling. A test shows that the paper responds to bleaching in the ordinary way, and a dark color can be lightened to almost any desired degree. Unfortunately, the color hue is so strong, and unrelieved by any whites, that it dominates almost any image that you can put on the paper, whether solarized or not. For this reason, we have found little use for colored printing paper in our work.

Another commercial product that, at first, showed great promise is Develochrome, produced by the FR Corporation. This is a set of chemicals, developer, and fixing bath that will produce a colored image instead of the normal black one on ordinary bromide paper. The hues available are red, blue, green, yellow, and brown. The chemicals come in powder form and are mixed and used like ordinary processing solutions. You must use the special fixer provided, because if you use ordinary fixer, the image color will turn to that of ordinary silver without a trace of hue. We were delighted with our first results, and we had a great time developing in Dektol and solarizing in Develochrome or vice versa, or developing in one hue and solarizing in another, producing two color images on regular bromide paper. The colors were not very saturated, but were pleasant enough for a nice variation from black-and-white photographs. Very soon, however, the results became less satisfactory, and we realized that the solutions were becoming exhausted. After about ten prints had passed through the developer, the colors began to come up more slowly and were less saturated. Even if the solutions were not used, their potency faded after about an hour or so in the tray. We went to replenish our supply, and found that the price per package had suddenly increased by a factor of three, thereby putting the price of Develochrome in the same class with the price of chemicals for regular color printing. Further disappointment came a few months later, when we found that the first of our Develochrome prints was already fading to a tinted pasty grey. To make an illustration for this book, we first developed the print in Dektol to get a strong silver image, then rinsed it thoroughly in water, placed it in a fresh tray of blue Develochrome, and turned on the light. As we watched, the background of the print gradually became a beautiful blue, and the figure obtained a brilliant white halo. We fixed the print in the special fixer and washed it. While it was still wet, we put it on the copy stand and recorded it on Ektachrome-X color film, thus ensuring it a greater measure of permanence. This is the picture on the cover of this book and on the first color page.

Develochrome does solarize well and predictably, and the colors are very pleasing. When they are developing, the solarized colors are very fragile and are easily marred by fingers or tongs. When removing the print from the tray, always hold it by a part that has already finished developing (the part that was exposed under the enlarger, not the solarized part). The paper that we used was Agfa Brovira No. 6. It is not on the manufacturer's list of recommended papers for Develochrome, and perhaps our problem with fading is a consequence of our not obeying instructions.

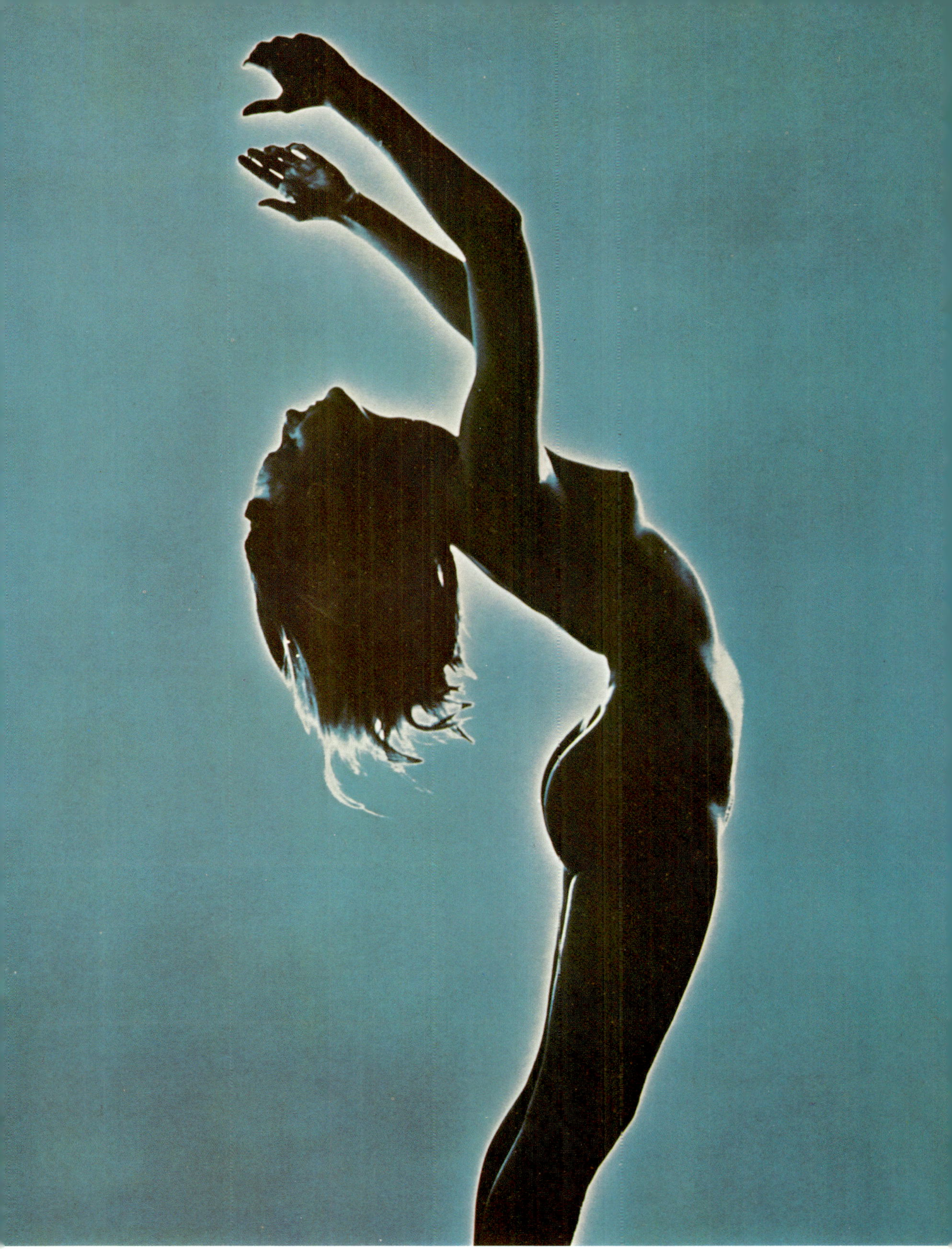

Plate 1

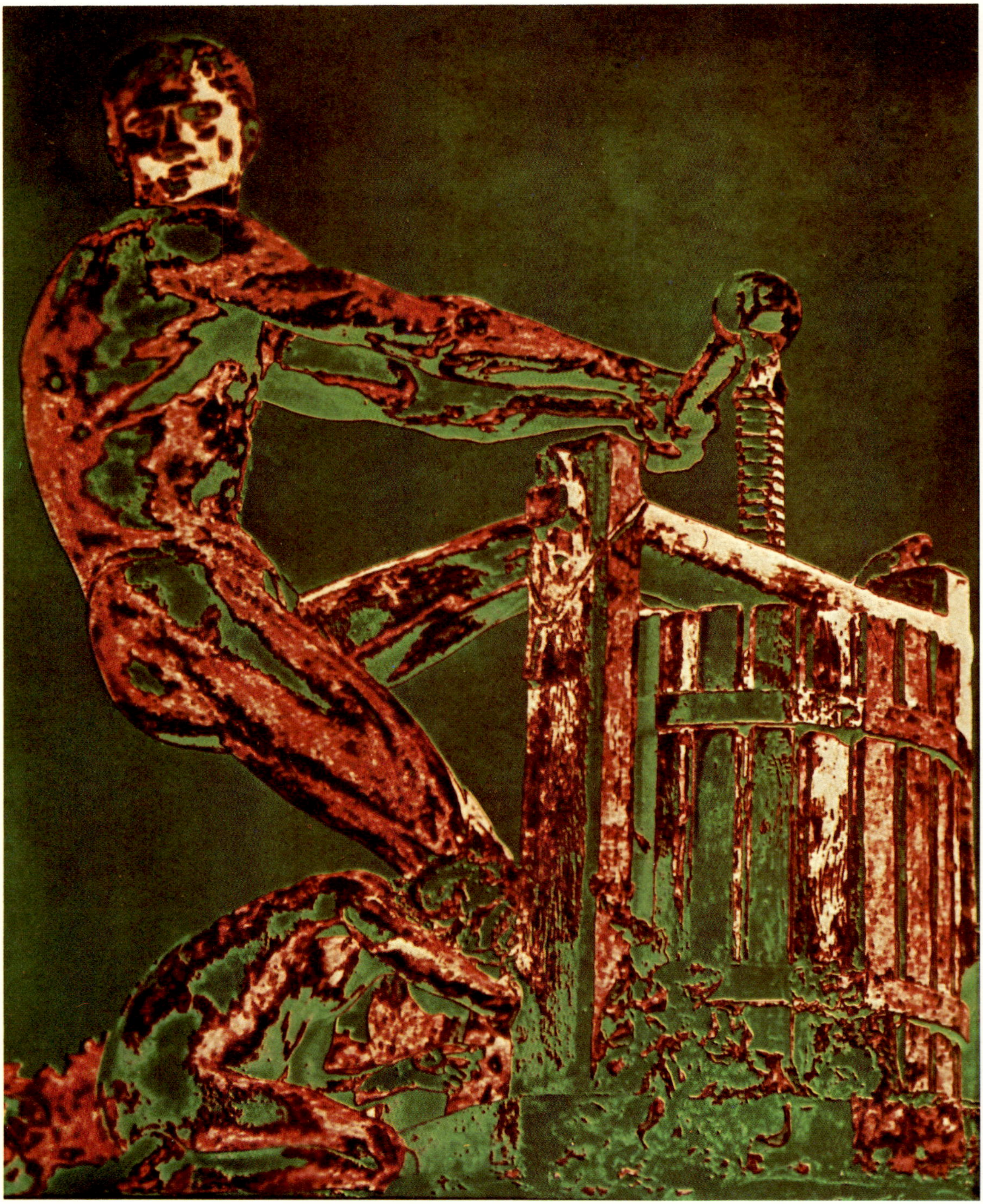

Plate 2

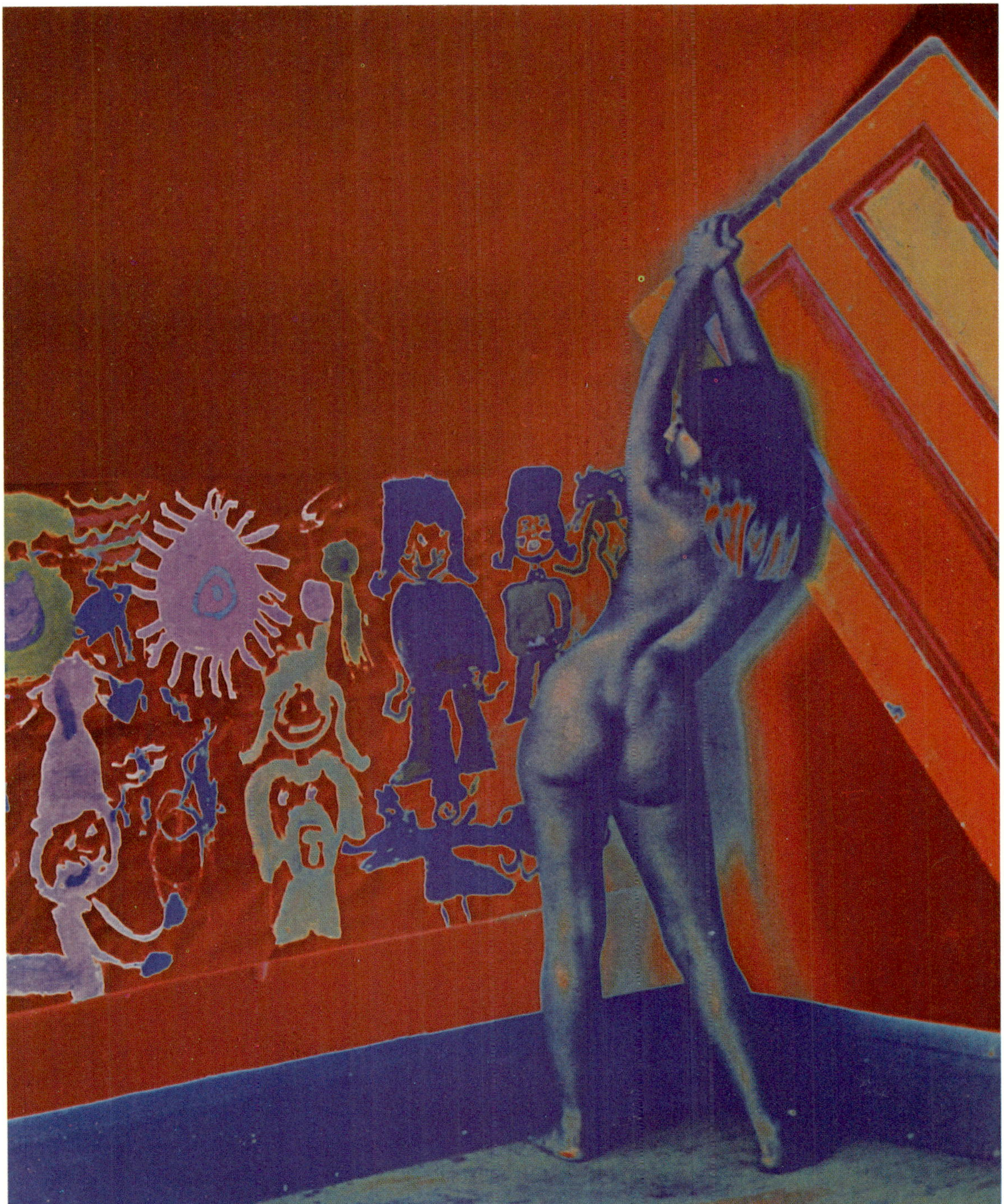

Plate 3

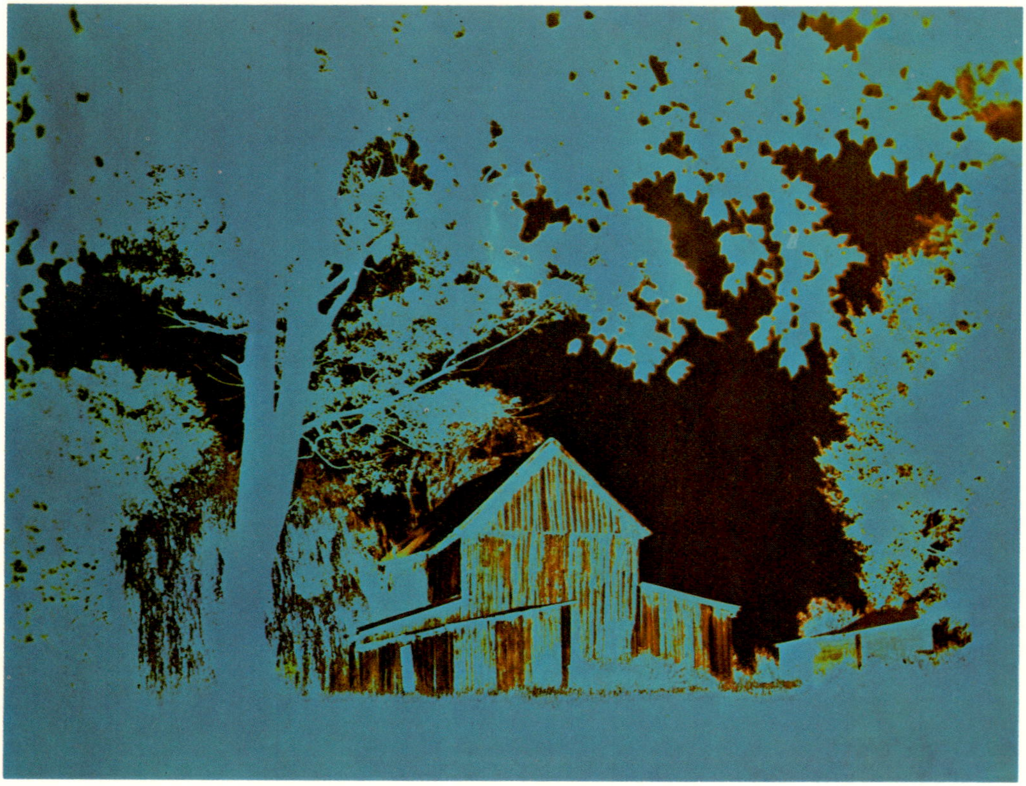

Plate 4

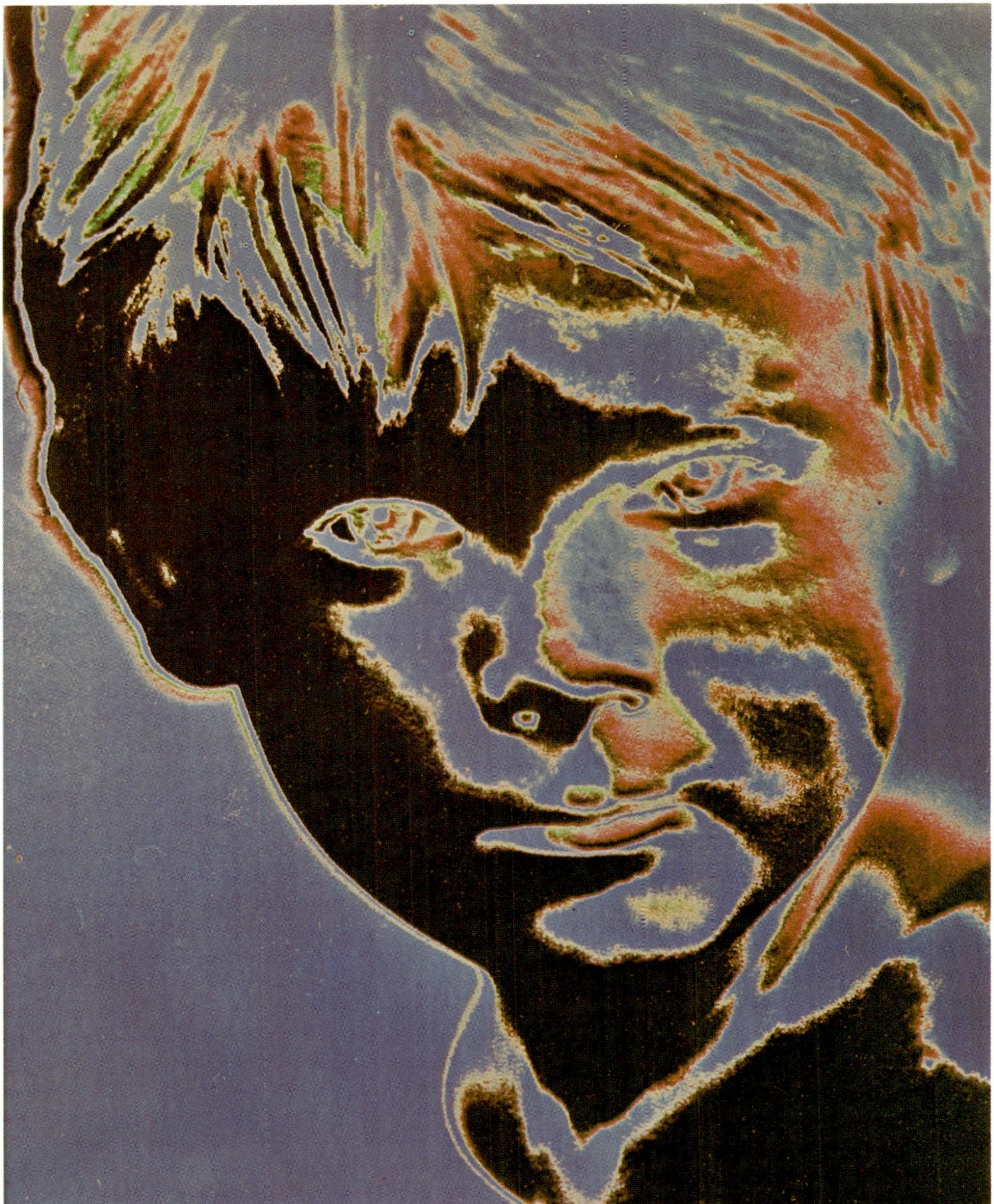

Plate 5

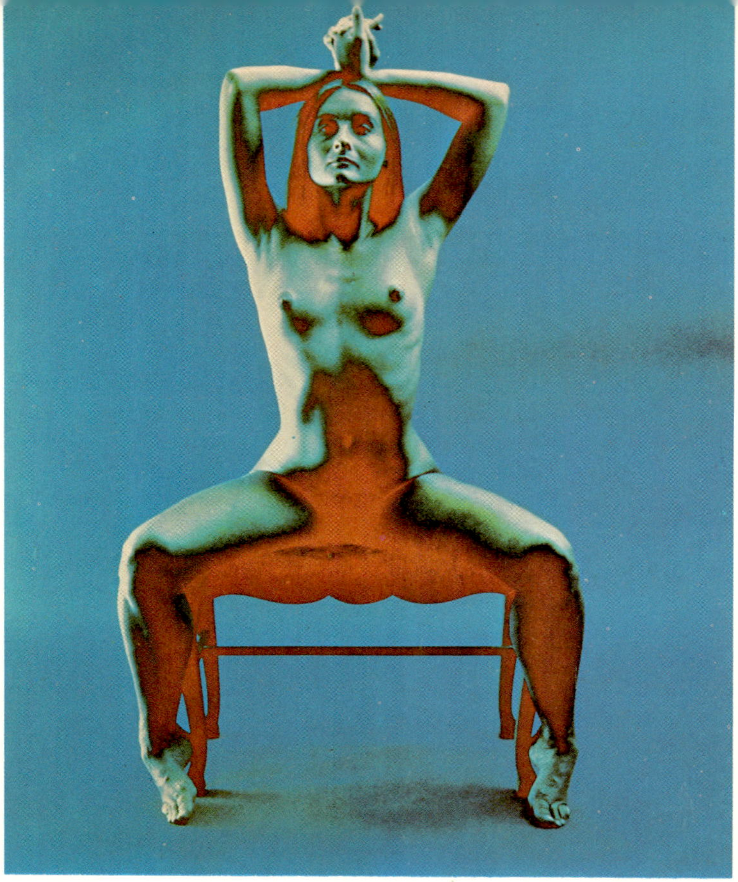

Plate 6

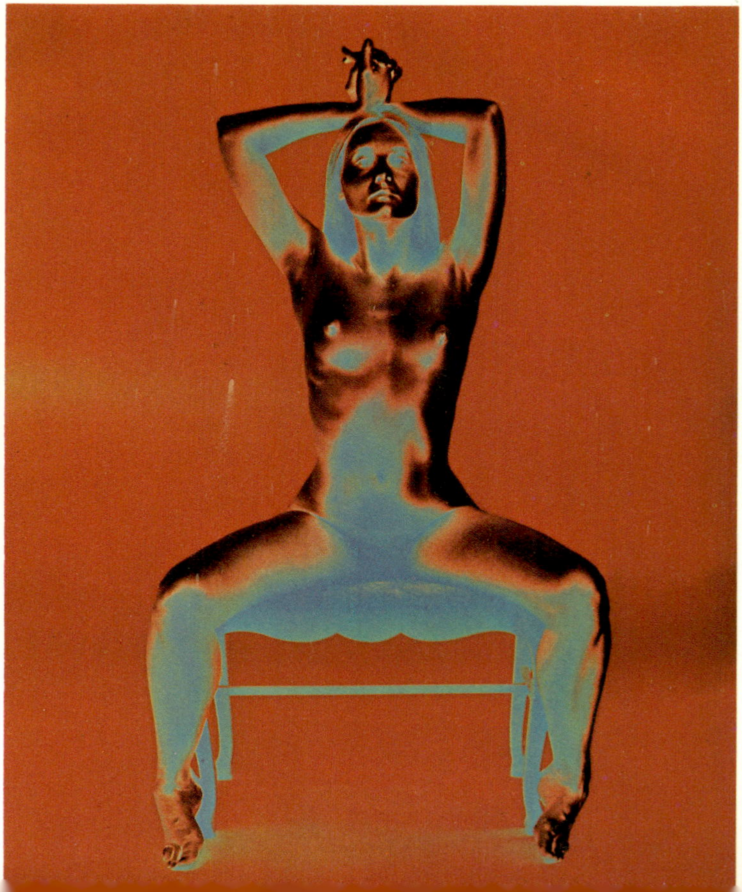

Plate 7

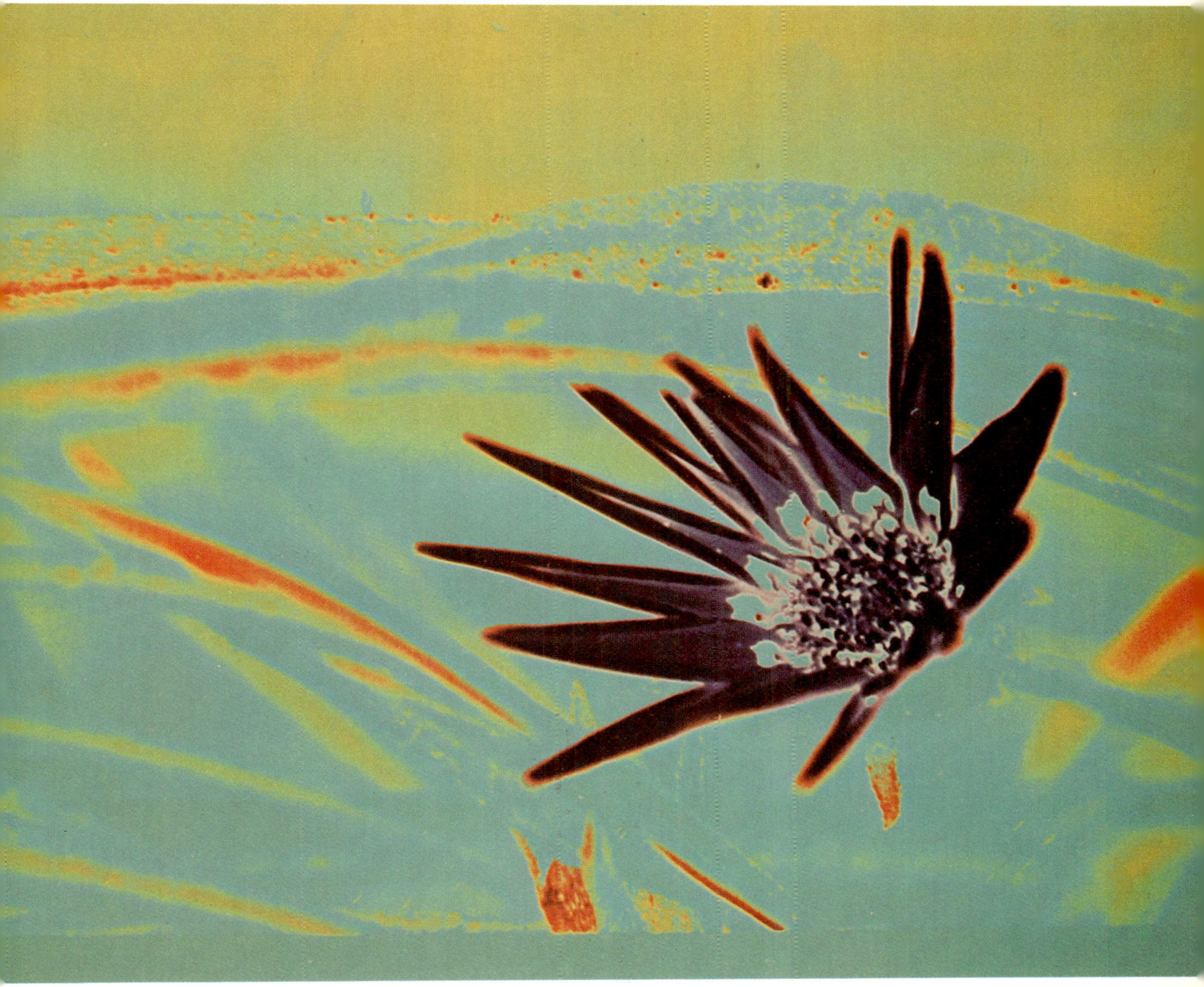

Plate 8

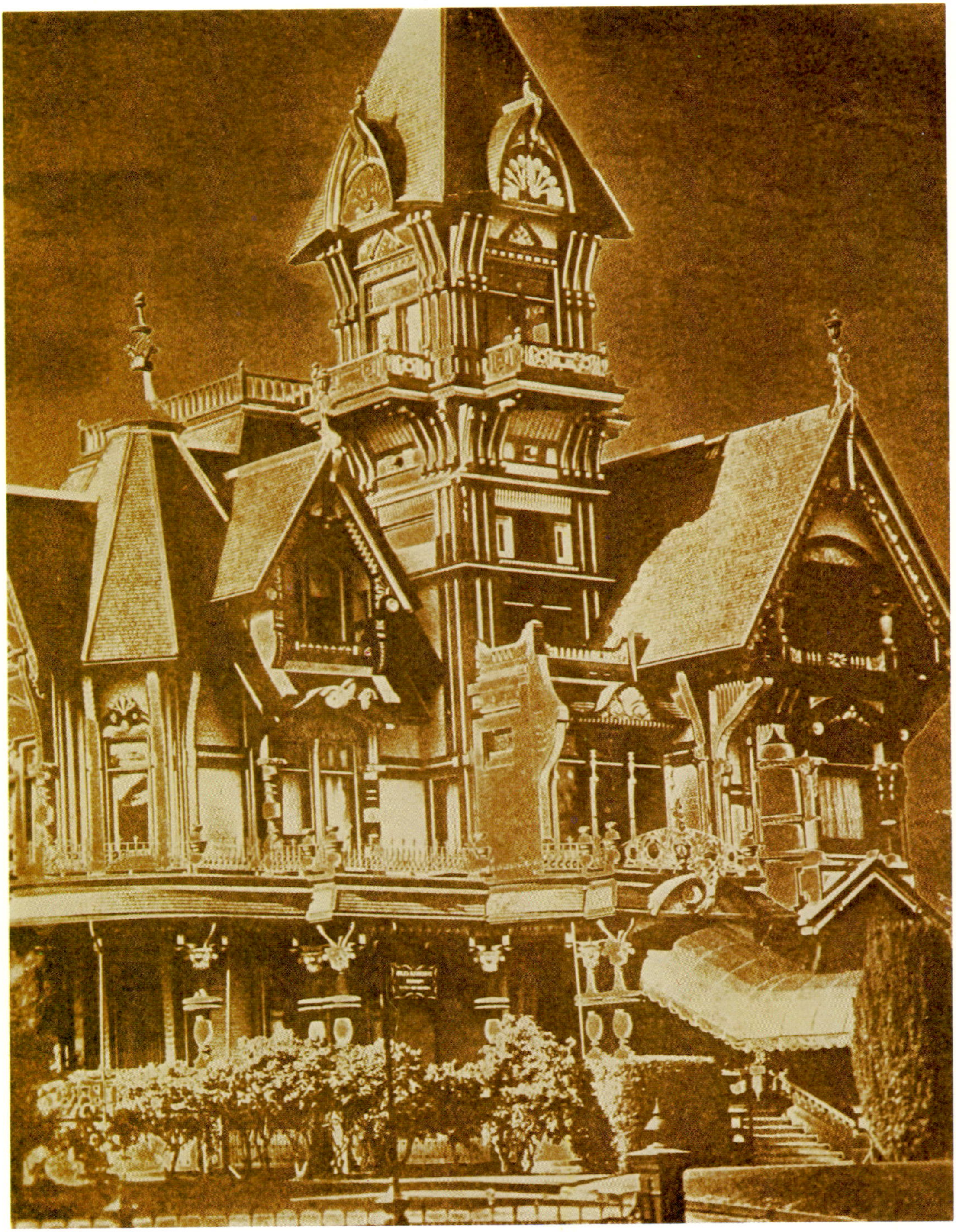

Plate 9

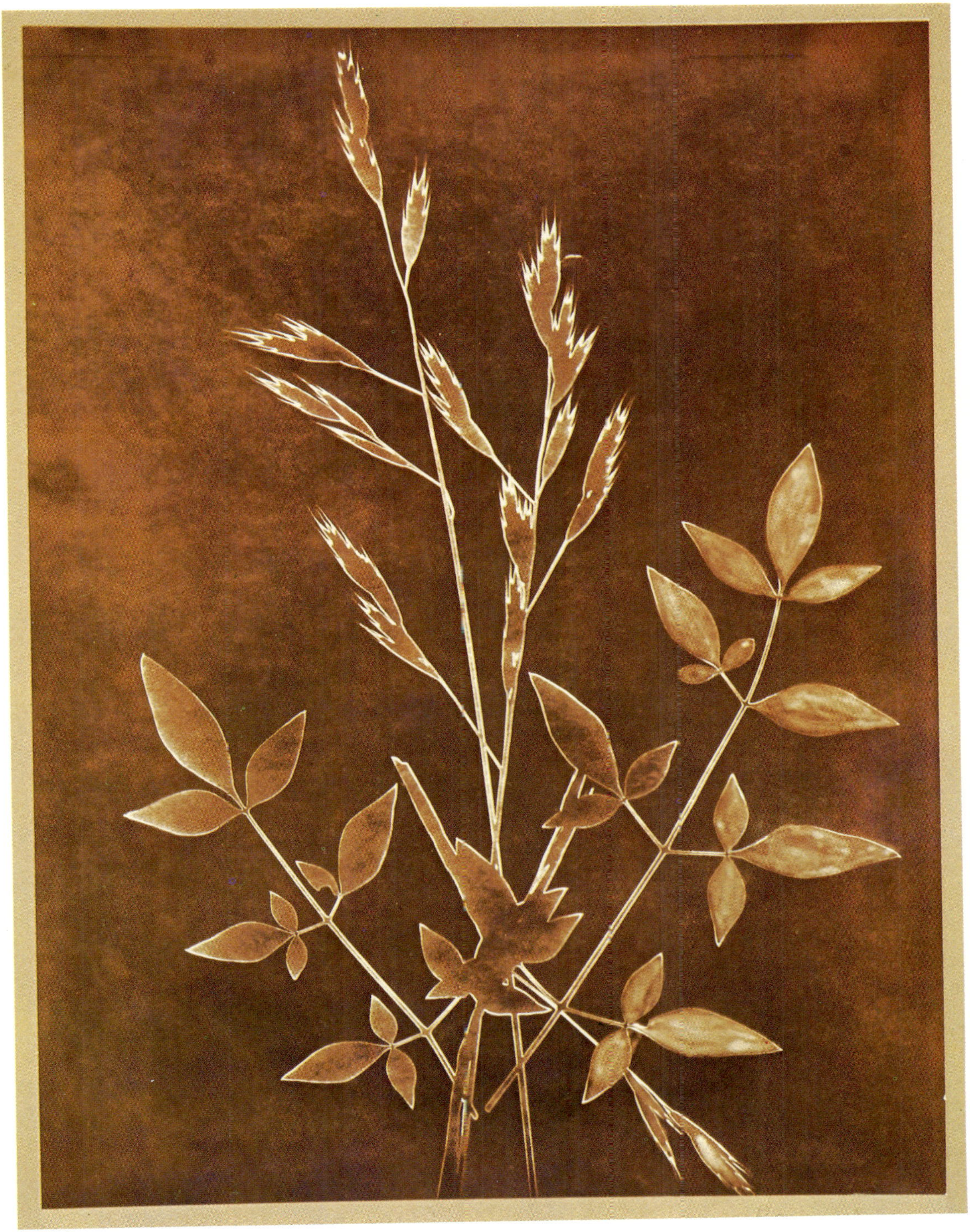

Plate 10

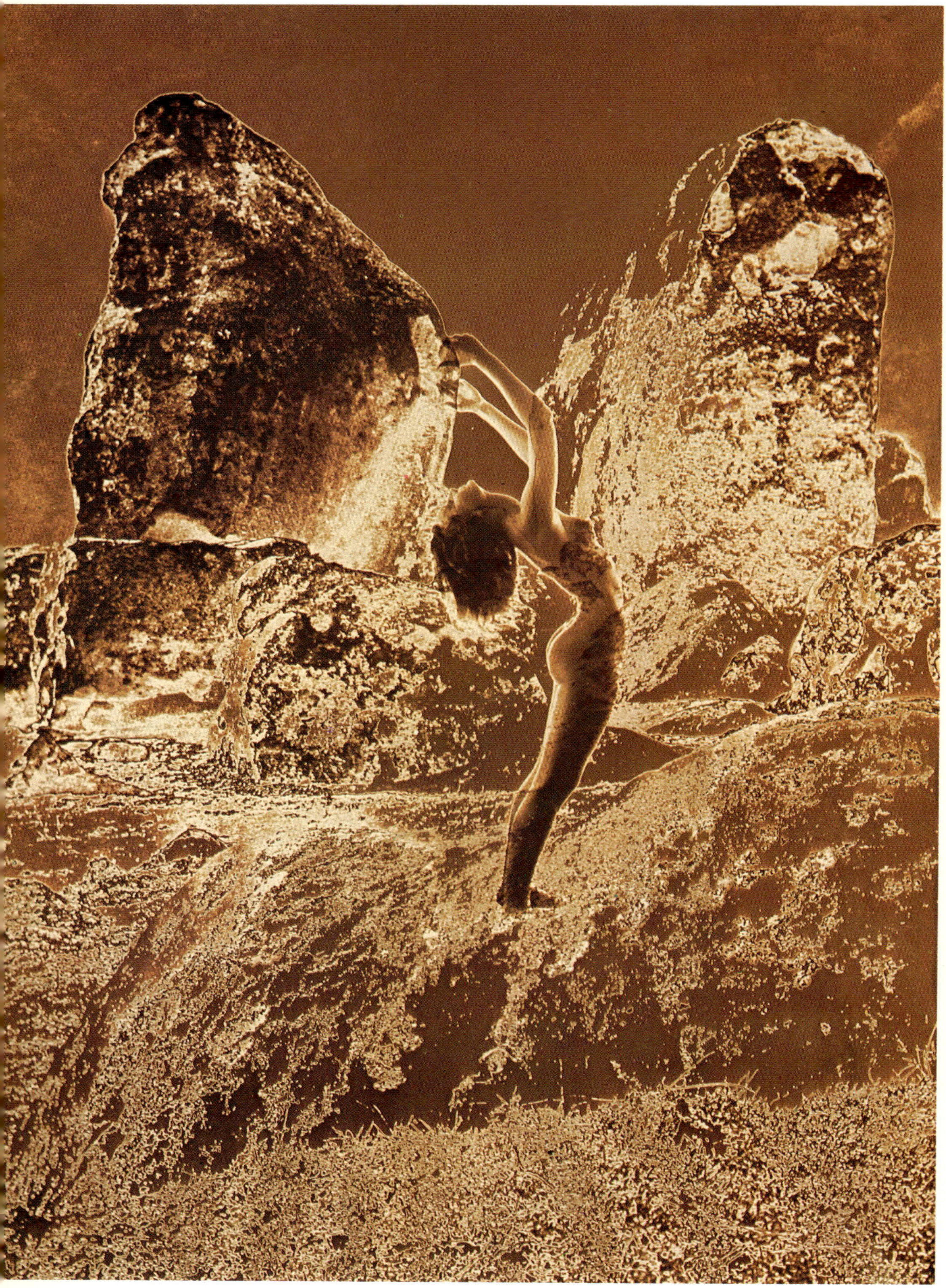

Plate 11

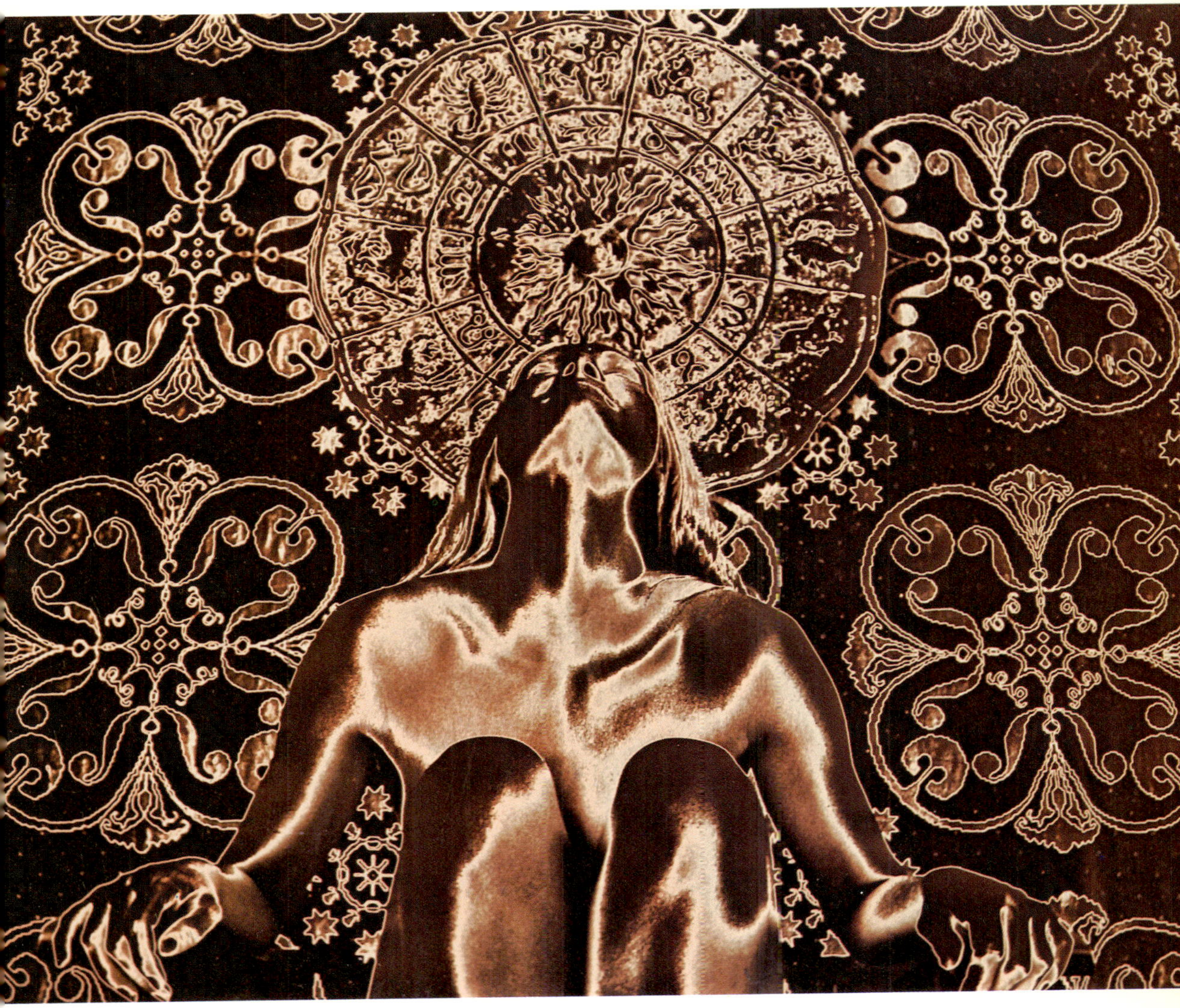

Plate 12

Plate 13

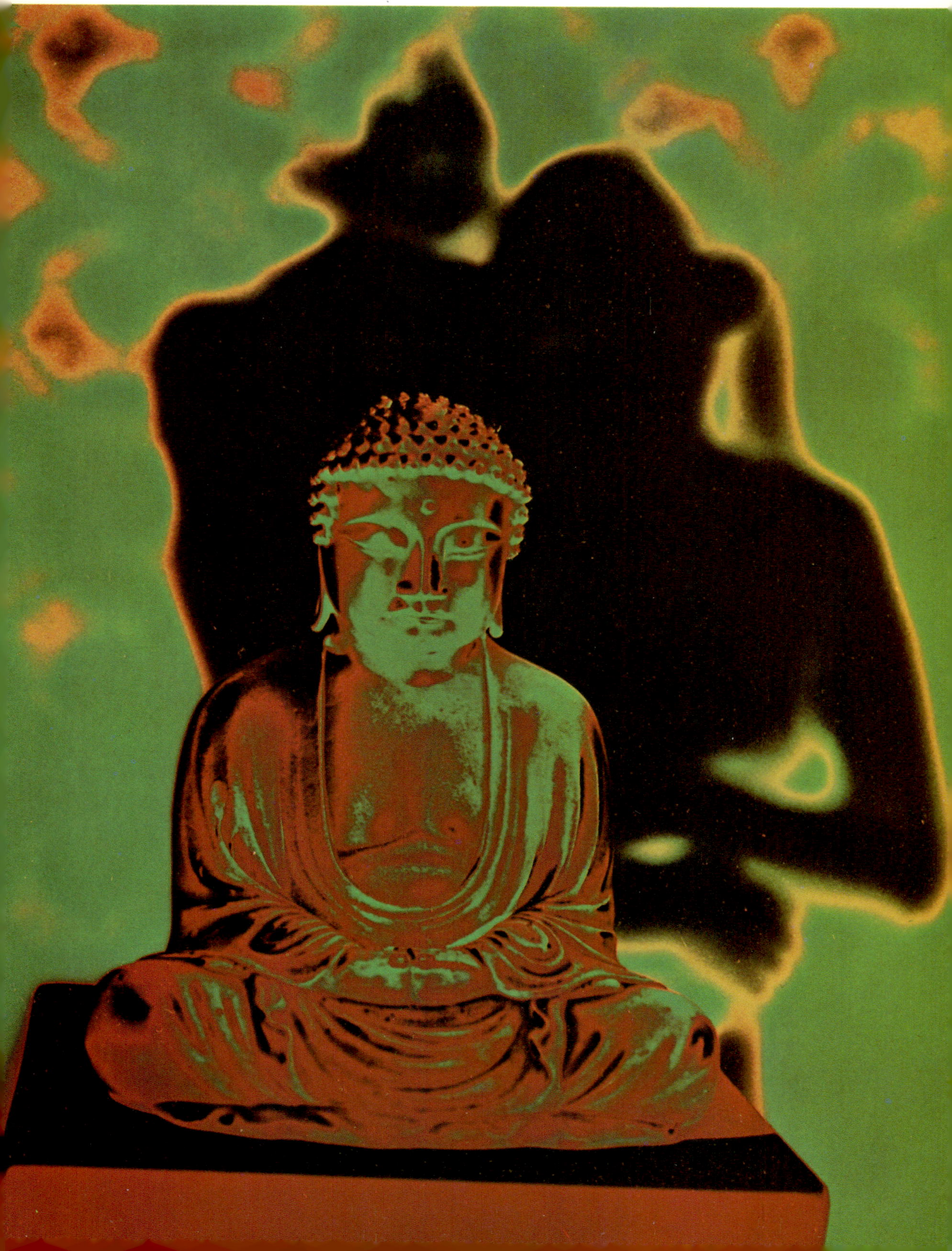

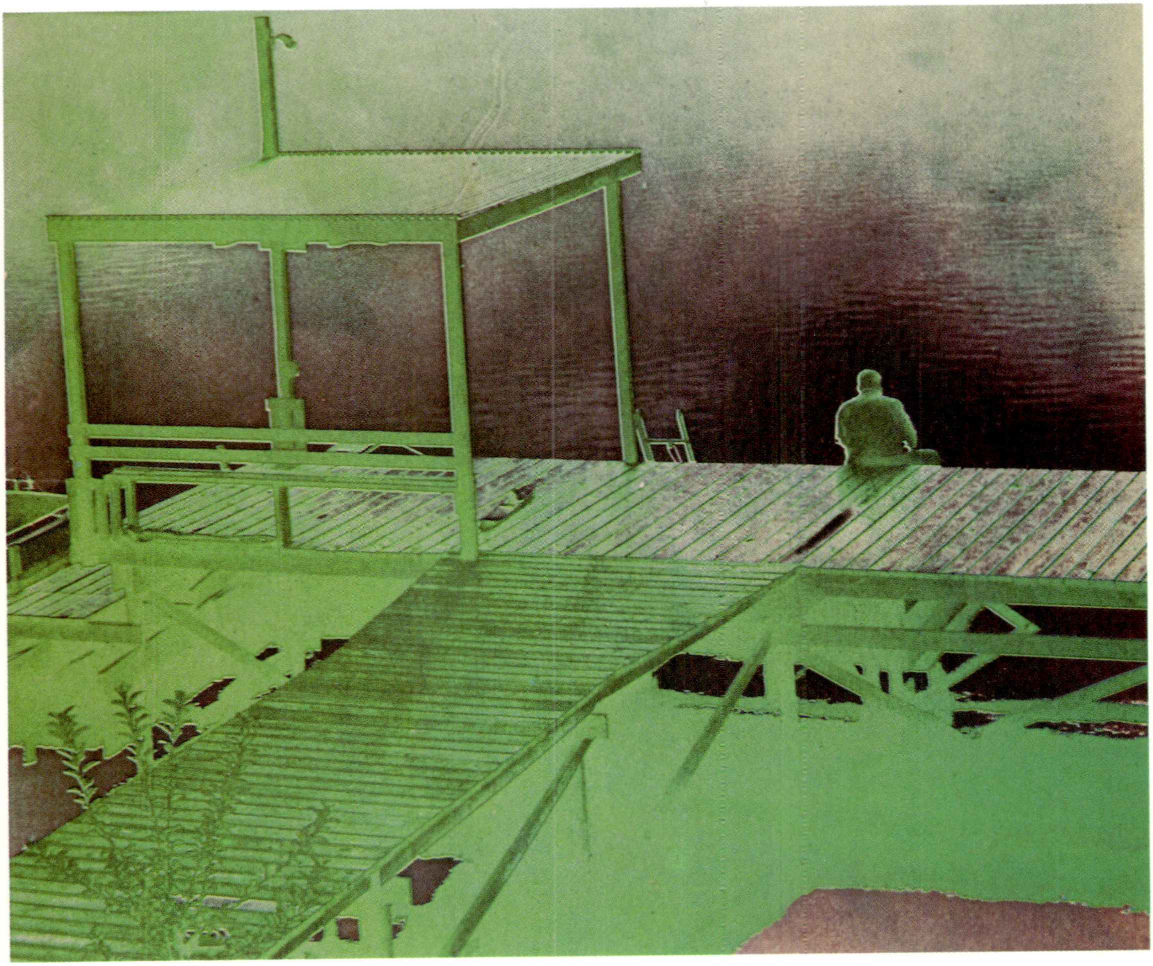

Plate 14

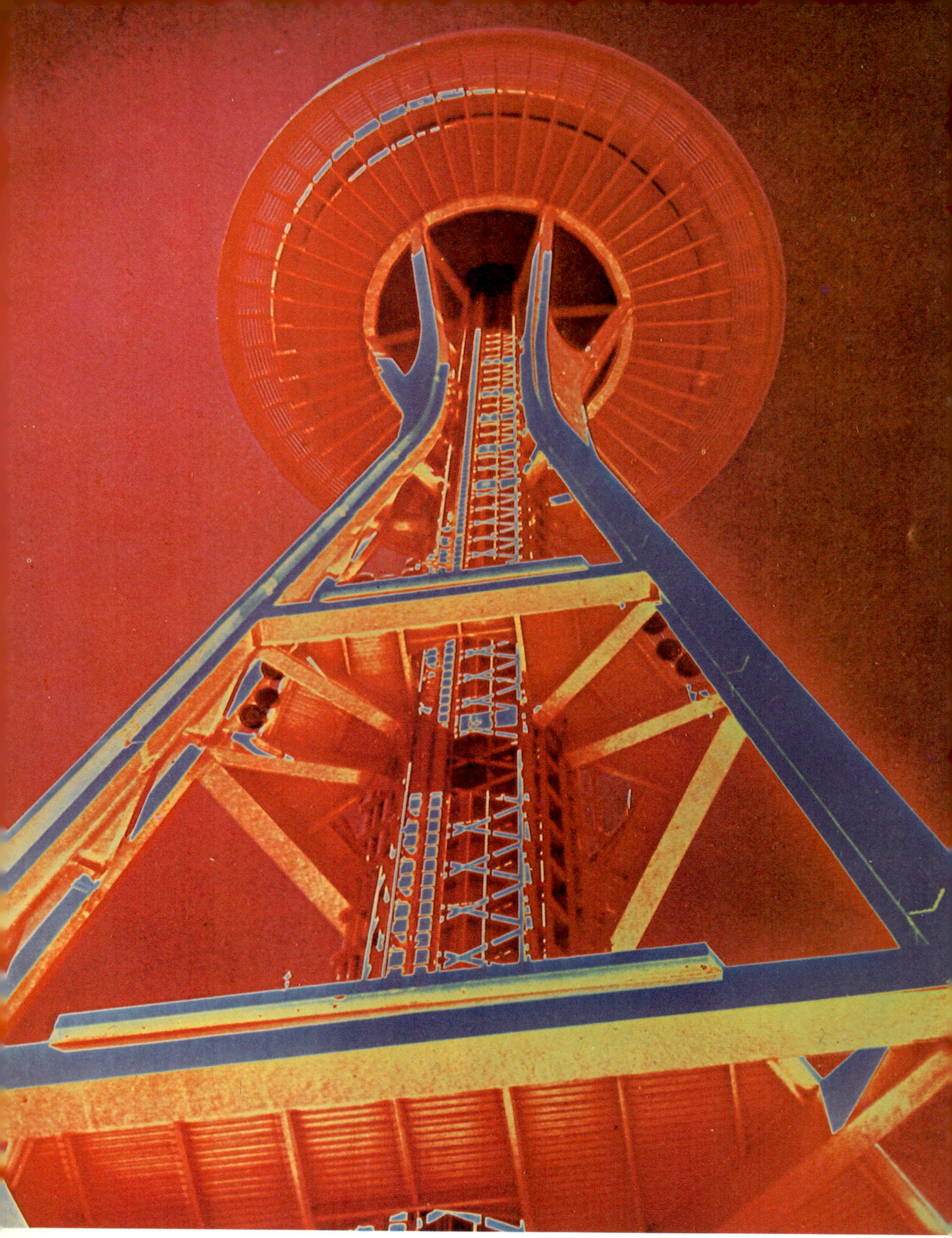

Plate 15

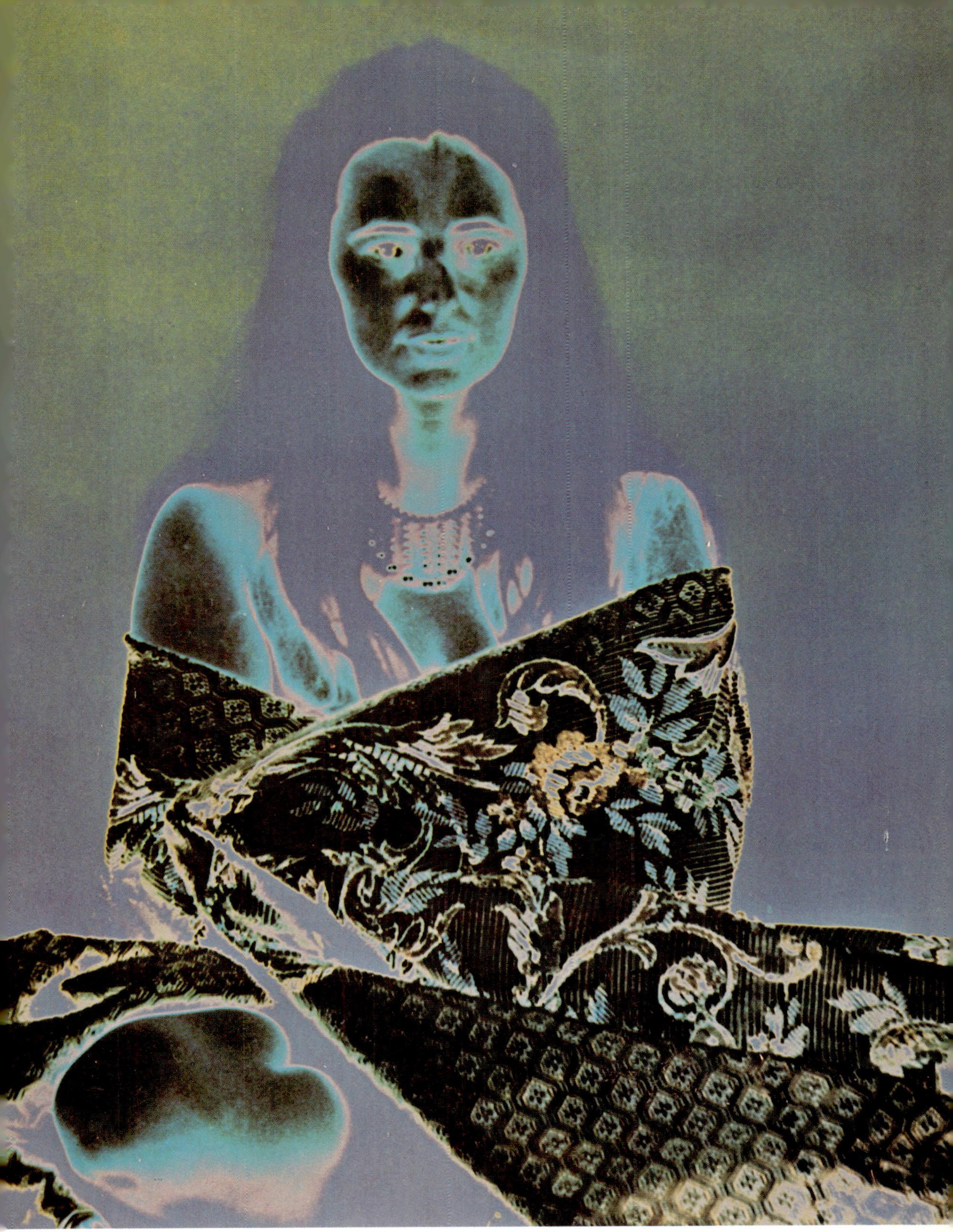

Plate 16

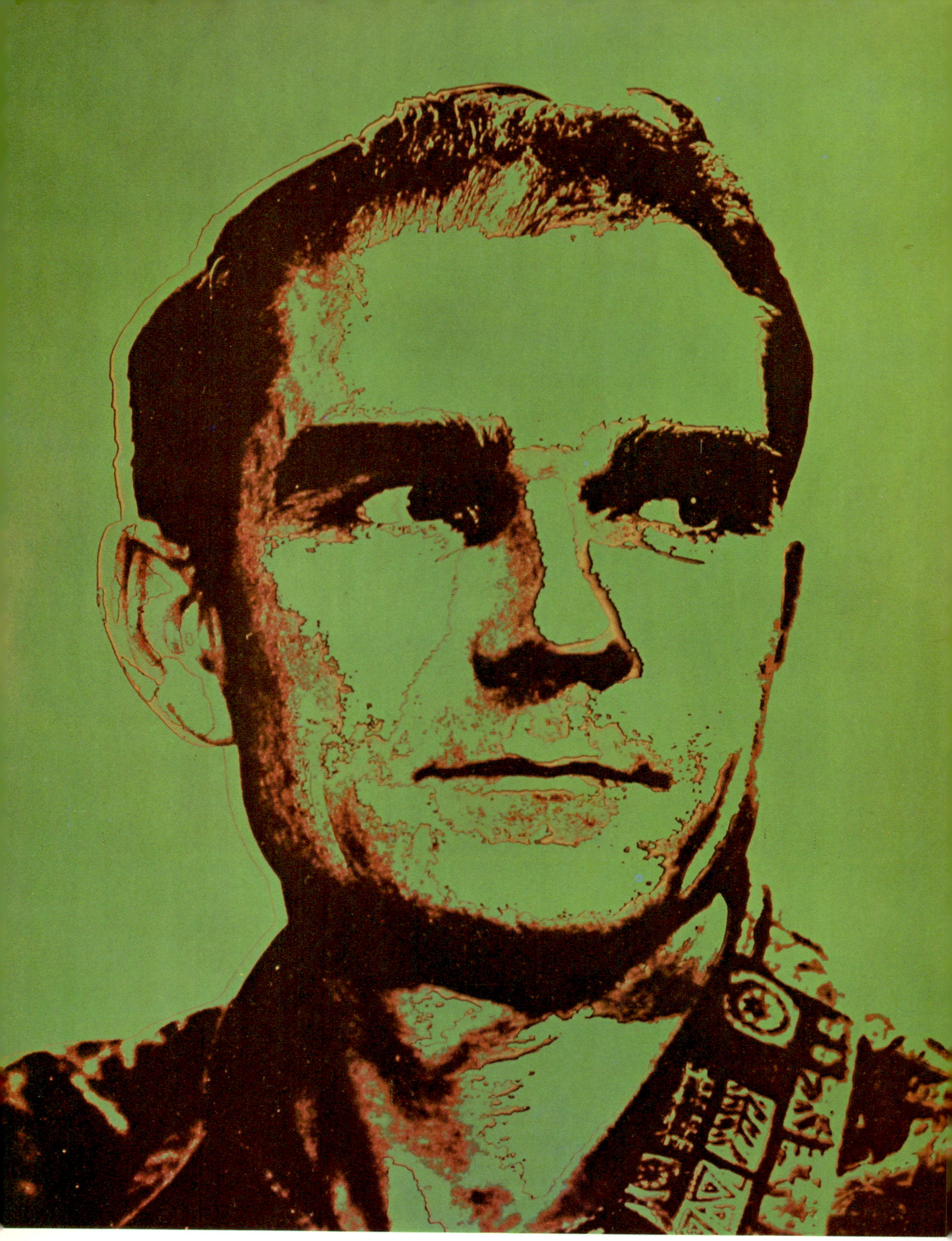

Plate 17

4

Solarizing Color Prints

Although many examples of solarized color pictures have appeared in print, very little seems to have been written about the process. Often, it is relegated to that vast land of color derivations where, it is suggested, you must muddle along as best you can; and it has been mentioned that solarizing a color print will usually yield nothing but mud. Note the word "usually." With sufficient knowledge of the materials and the process, it is possible to avoid the usual result and achieve very good solarized color prints. A thorough knowledge of the science of color is not necessary, but it is helpful to know a few basic facts. The Kodak publication, *Color As Seen and Photographed*, is an excellent introduction to color principles as they apply to photography, and is the basic primer for anyone who is serious about making pictures in color.

To most people, the word "color" brings to mind the visual sensations by which we distinguish the various hues of the spectrum—red, orange, yellow, green, blue, and violet. There are, of course, many intermediate hues, such as blue-green, or orange-red, and so on, as well as many tints and tones, representing the various degrees of saturation and brightness of the colors. The remarkable fact of human vision that makes color photography possible is that all of these sensations can be produced by adding varying amounts of light of only three colors. These three hues, known as primaries, can be chosen in many ways, the only requirement being that when all three are present in the proper proportions the result appears as white light. The colors usually chosen as the primaries are red, blue, and green, which correspond roughly to the sensitivity regions of the three types of color-sensitive cells in the retina of the eye. For our discussion, then, it is sufficient to consider the spectrum of white light as consisting of three parts—red, blue, and green.

Projecting separate beams of light of different hue onto the same area is called additive color mixing, a very important process in the scientific study of color. Modern color photography depends more on subtractive color mixing, which involves passing originally white light successively through colored filters, each of which removes one hue from the beam. A filter (or a dye) that absorbs blue light from the beam but transmits all the rest is yellow in appearance, and is called minus blue (complementary to blue). This minus-blue light, which contains rays from all parts of the spectrum except the blue, may appear to be exactly the same hue as a beam of yellow light, which contains no other rays except yellow. Although the human eye is incapable of distinguishing minus-blue from pure yellow, the color emulsion may respond quite differently to them.

Likewise, a filter or dye that absorbs only green light will transmit both red and blue and appear purple, or magenta. Magenta is thus known as minus-green. A filter that absorbs only red will transmit the green and blue, giving the color known as cyan, or minus-red. These three colors, yellow, magenta, and cyan, are called subtractive primaries, and are the colors of the dye layers in the tripack emulsion of color printing materials. The amount of each dye formed in the emulsion is determined by the amount of its complementary light in the exposure. For example, exposure to blue light causes the formation of yellow dye upon development. The cyan and magenta dyes are formed during development, after exposure to red and green light respectively.

The emulsion coated on the printing paper consists of four separate layers. The top layer is sensitive only to blue light and contains molecules that form a yellow dye upon development. The second layer is a yellow filter that prevents the blue light from passing through and striking the two lower layers, which are also sensitive to blue light. The yellow filter is bleached in processing. The layer beneath the yellow filter is sensitive to green light and contains molecules that form a magenta dye upon development. The bottom layer is sensitive to red light and contains molecules that form a cyan dye. Because the dyes are not perfect (they absorb some light that they should ideally transmit), most negative color films have an orange hue that serves as a mask to correct some of the imperfections. It would seem reasonable to have color print paper balanced to produce a neutral grey or black when exposed to white light through the orange film base. Unfortunately, this ideal is seldom achieved in practice and the proper color of the enlarger light must be experimentally adjusted by the use of filters, usually placed between the lamp bulb and the condensers.

Gelatin filters, designated as CP for color printing, are available in graded densities of the three subtractive primaries, yellow, cyan, and magenta, and are calibrated so that they may be used in combinations to produce any required shift in the color balance of the enlarger light. The use of these filters in normal color printing is fully explained in the *Kodak Color Dataguide*, which also contains viewing samples, a filter computing wheel, and other practical aids that are useful whether or not you are using Kodak's color process. If color printing is new to you, we recommend that you experiment with it a bit before venturing into color solarization. This does not mean that you must become an expert in the production of beautifully balanced color prints, but only that you be familiar with the filter combinations and the color shifts that they can produce in your prints. For derivation work, it is a good idea to start with a filter pack that will produce a neutral grey on your printing paper when there is no negative in the enlarger carrier.

For our work with color solarization, we used the Unicolor three-bath process because it is simple and quick. At least two other manufacturers produce similar systems that should work just as well, and the procedures can be modified to suit the particular chemistry you find most convenient. Our methods were adapted from those we had used with black-and-white prints; after the development was two-thirds completed, the print was quickly rinsed and transferred to a tray of water for the second exposure. The light used for the second exposure was provided by a 15-watt bulb in a safelight fixture. Instead of the regular safelight filter, a selected pack of CP filters was placed in the holder located two feet above the tray. Depending upon the color and density of the filter pack, the second exposures varied from two to about fifteen seconds. The print was then placed back in the developer for the remaining third of the required developing time. Then came a rinse, a bleach-fix, another rinse, a stabilizer, and the print was ready to dry. The total dark-time required was three and one-half minutes, broken by the brief seconds of the solarizing exposure. An old metronome, set to beat seconds, made an excellent audible timer.

Our experiments with color solarization were done out of curiosity, with no intention to carry out the extensive testing that we had done with black-and-white emulsions. That work had been initiated by the discovery of "old brown," and we made no attempt to find its equivalent in color processing. It actually was not necessary to do so, because the dye images showed much less of a tendency to general fog than the silver image processed in fresh developer. This was also true of Develochrome. In all cases, of course, we are dealing with the photo-reaction of silver halide, the dye image being substituted for the silver one after the solarization has taken place.

Although we did not investigate the comparative solarizing properties of used and fresh developer in color processing, we did discover that the chemicals could be used well beyond the manufacturer's recommended lifetime. In fact, we processed over three times the recommended number of prints per gallon of solution without any significant deterioration in the results. It is quite possible, however, that had we been much concerned with accurate color balance and working speed, such overuse of the chemicals would have been inadvisable. We adhered to the recommended working temperature of 85° F. for the solutions, but found that the tolerance was not critical for our purposes; it should be possible to work just as well at a fixed temperature as much as five degrees more or less than recommended, provided processing times are adjusted accordingly.

The first and second exposures in color solarization are nearly always of entirely different hue. This provides a challenge to the sense of color combination and harmony, particularly since the two exposures give not only their own complementary hues, but also mixtures of the two. We tried out a number of colored light sources, including safelights, plastic and gelatin theatrical filters, and colored light bulbs. The yellow OA safelight gave a brilliant blue when used for the second exposure, and some of the theatrical gelatins provided pleasant surprises, but we found it impossible to predict the results with them because we did not know their true spectral transmission.

Only with the graded CP filters could we reasonably guess what kind of results a particular combination would give. Even then the amount and hue of the first exposure had to be considered, and in spite of careful thought beforehand, we were often surprised. With the CP filters, it was not difficult to shift the colors in any desired direction, using the simple rule: To increase the amount of a hue in the print, decrease the amount of that hue in the filter pack. This rule, of course, holds as well for the filter pack of the second exposure as it does for the first.

Since we were not at all concerned with true color balance, but only with a pleasing result, the real problem was achieving a harmonious combination. We found that the purest hues were produced when the original was a positive color transparency rather than an orange-masked color negative, because the printing contrast was much higher. When working with high contrast originals, the first exposure and development would give a colored image with large areas of pure white. The second exposure and development would then color the white areas with the complement of the exposing light, and yield intermediate hues in the edge effects while affecting the original color very little. Low contrast originals, on the other hand, would give no completely white areas on first development, and the combination with the second exposure color was likely to be an unsaturated hue that was often unpleasant.

The effect of various combinations was particularly difficult to predict because of the layered structure of the color emulsion, and depended very strongly upon which exposure was given first, its color, strength, and development. For instance, if the first exposure was red, it would affect only the bottom layer of the emulsion, so the second exposure would be free to act on the top two layers. If, however, the first exposure was blue, the first development would produce density in the top layer, thus partly shielding the bottom two layers from the action of the second exposure. It turns out that experimentation is the only good way to find out what will happen in any particular case.

It isn't necessary to start with a colored original in order to produce a color print. Black-and-white solarized prints can be used as masks along with black-and-white negatives either on paper or on film.

The off-beat tonal values in the solarized images make solarization ideal for introducing unusual color effects into pictures originally taken in monochrome. Color Plates 1, 2, 5, 9, 10, 11, 12, and 17 were all derived from monochrome originals. All the other color plates were created from straight color transparencies. Color printing with black-and-white masks involves problems of registration just like those discussed in relation to tone separation, and the procedures are just the same. Any positive or negative print can be used as a paper negative and printed on color paper, the exposure being made with a colored light bulb or through a theatrical gelatin. The paper is not developed, but a second exposure is made with a different colored light using a solarized version of the same paper negative in exact register with the first image. This twice exposed print can be given a solarizing exposure to a third color while developing, if you want to take the derivation one step further. The scarcity of detailed information on color solarization is quite understandable. The number of possible variations and combinations of the color and intensity of the exposing lights and different processing conditions is almost astronomical, and any written instructions can encompass but a minute fraction of them. The description itself is more complicated and much less informative than actual experience in the darkroom. Only by actually working with color do you become familiar with its possibilities and limitations. You will probably find it easier than you expected, and much more fun.

5

The Science Behind the Art

Aside from the required aging of the developer, there are four principal variables involved in the control of the Sabattier effect. These are: the first exposure, the first development, the second exposure, and the second development. Other variables of secondary importance are: the colors and intensities of the exposing lights, the temperature of the developer, and the agitation of the print during development. Still other factors are such things as the depth and color of the developer, and whether the print is removed from the developer and washed before the second exposure is given. Also of importance are the type, brand, and contrast of the printing paper you are using.

As with any other complex process, standardization is required in order to achieve consistency. If, in addition, you want to be able to produce any desired result, it is necessary that you know the effect of each of the variables on the finished product. Obviously, given the large number of variables, this is not an easy job, and when it is realized that the combined effect of two variables is nonlinear (that is, not simply additive), then the task begins to look impossible. It is this very complexity, however, that allows us such a great amount of control, and that makes the process so versatile. Needless to say, our present study is far from exhaustive. We have, in most cases, accepted convenient properties and values as standard for most of the variables, and deliberately changed only a few to achieve our results in the simplest way. It is our purpose in this chapter to set down some thoughts and observations concerning the solarization process, and to report some results of various experiments we have performed in the past two years.

Of the four principal variables mentioned above, the first exposure and the first development are the most critical. Roughly speaking, we might say that the first exposure determines the amount of detail that will be visible in the final print, while the first development controls the degree of reversal and the contrast of the image. Just as in normal printing, these two variables are interdependent, and a decrease in one can partly compensate for an increase in the other. Figs. 5-1A to D show prints (of a statue) that were processed identically except that they received different exposures in printing. We see that Fig. 5-1A is drastically underexposed, and consequently lacks detail and appears two-dimensional. The lack of texture is evident in areas where no image had developed before the solarizing exposure was made. Fig. 5-1B received what would be considered a "normal" first exposure, and texture and detail are near maximum. Overexposure, as in Fig. 5-1C, brings out further detail in the highlight areas, but the shadows have gone too dark and there is little reversal. Extreme overexposure (Fig. 5-1D), produces an almost positive silhouette with very little detail. Note that the sky becomes increasingly lighter with increasing first exposure. If you wish the sky to be black in a solarized print, you must expose so that the first development will produce no density in that area. The bright patch at the upper left of Fig. 5-1C shows that this area was just beginning to darken when the solarizing exposure was made. In Fig. 5-1D, the first development had produced an appreciable density in the same spot and the middle region of the sky was beginning to appear. The bright halo behind the three figures in Fig. 5-1D is due to the inhibiting effect of the developer reaction products that have diffused into the area from the rapid development of the three dark figures. The building on the far left that is not visible in Fig. 5-1A, and is negative in Figs. 5-1B and 5-1C, has become positive in Fig. 5-1D.

Changing the first development time with a high contrast negative, like that of the statue, mainly serves to change the contrast of the solarized print. A shorter time means a darker print of lower contrast. If we begin with a

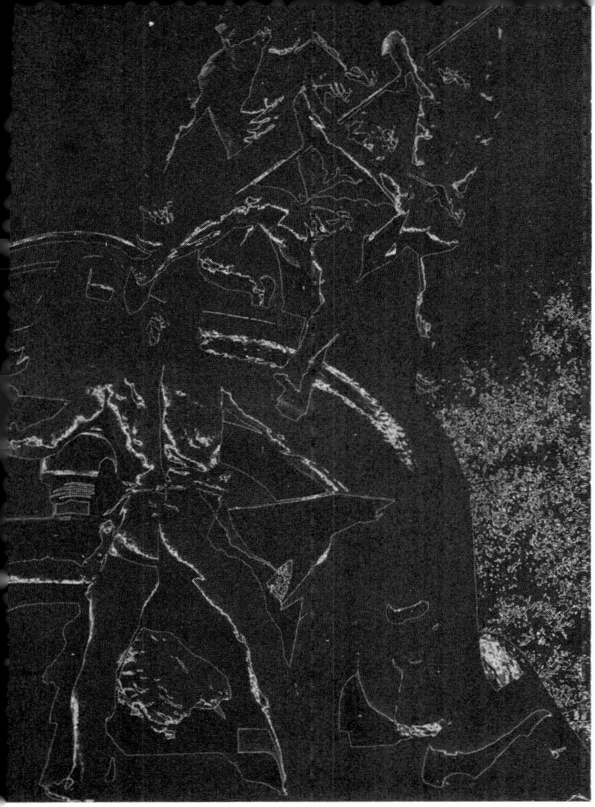

5-1A

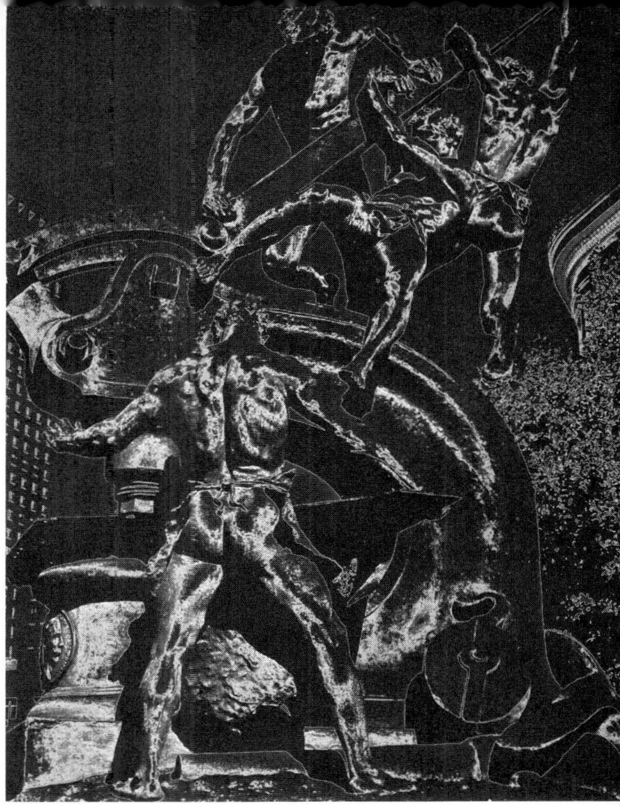

5-1B

5-1C

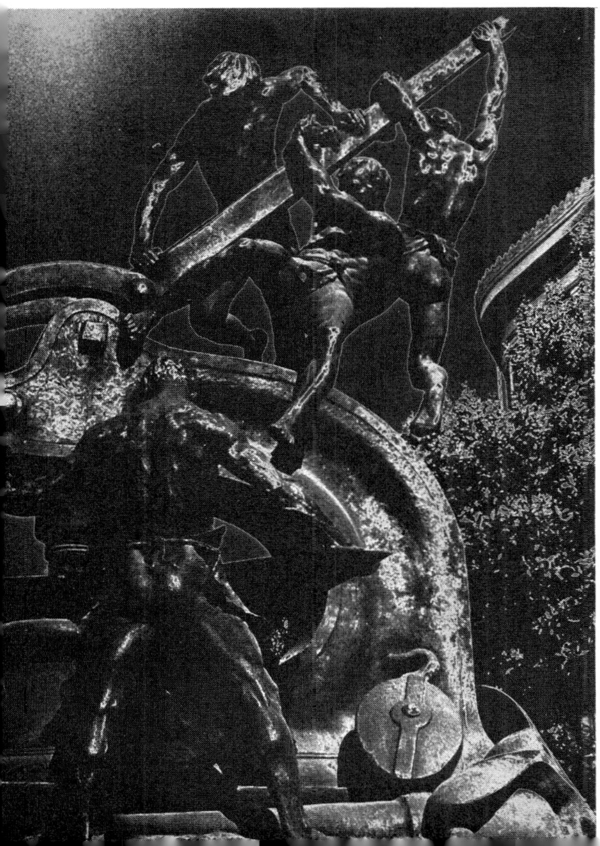

5-1D

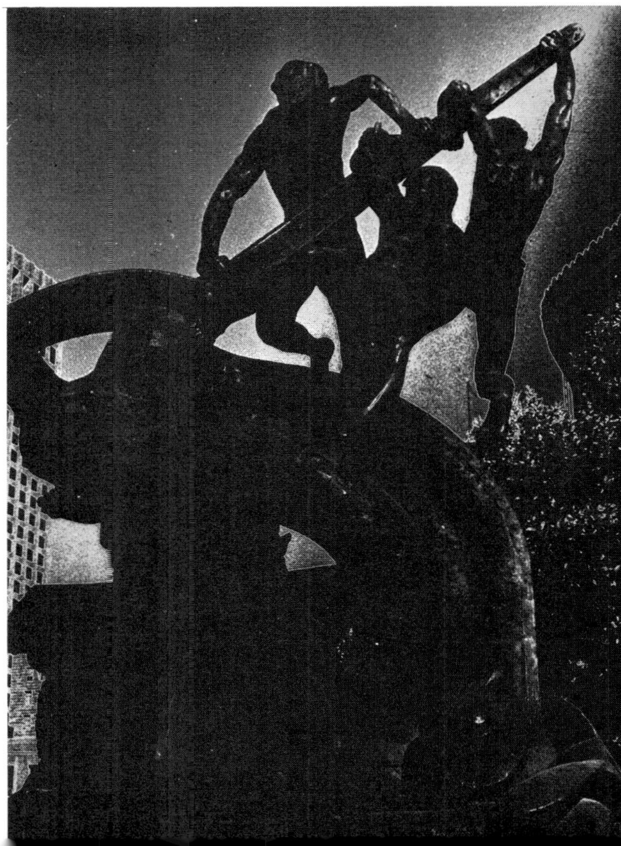

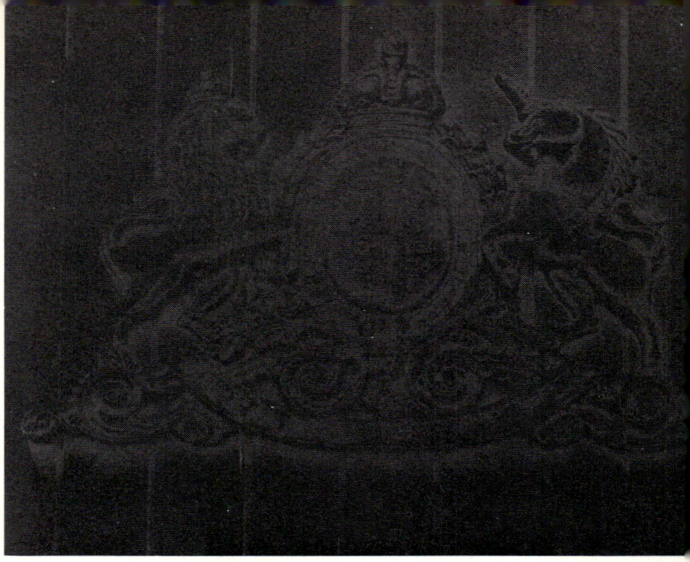

5-2A

low contrast subject, the first development can drastically affect the appearance of the resulting image, changing its character from positive to negative. The prints in Figs. 5-2A to G show some of the possible variations. The first four prints, Figs. 5-2A, B, C, and D, all received the same treatment except for the duration of the first development, which varied from 30 seconds for Fig. 5-2A to 90 seconds for Fig. 5-2D. The total time of development in each case was 3-1/2 minutes, the first exposure was 20 seconds, while the second exposure was 15 seconds to a 25-watt bulb 5 feet above the tray. The image in Fig. 5-2A was just barely visible when the second exposure was begun, and we see it is very dark and completely negative. Fig. 5-2B was developed 45 seconds before the second exposure, and is about half positive. With a 60-second first development, Fig. 5-2C is not much different from 5-2D, being largely positive except for the main shadow and some residual highlights on the unicorn. Figs. 5-2E and F were given the same treatment as Fig. 5-2D except that their first exposures were 12 seconds and 30 seconds, respectively. Fig. 5-2E is like a very high contrast version of 5-2A, while Fig. 5-2F is simply a dark, straight print, since the image was so fully developed that the second exposure had almost no effect. In fact, you can compare the other prints with Fig. 5-2F in order to see the changes produced by the solarization. This lion and unicorn emblem was photographed on the front of a restaurant in San Francisco. The upper right of the emblem was in shadow, while weak sunlight struck the lower left. The 35mm

5-2B

5-2C

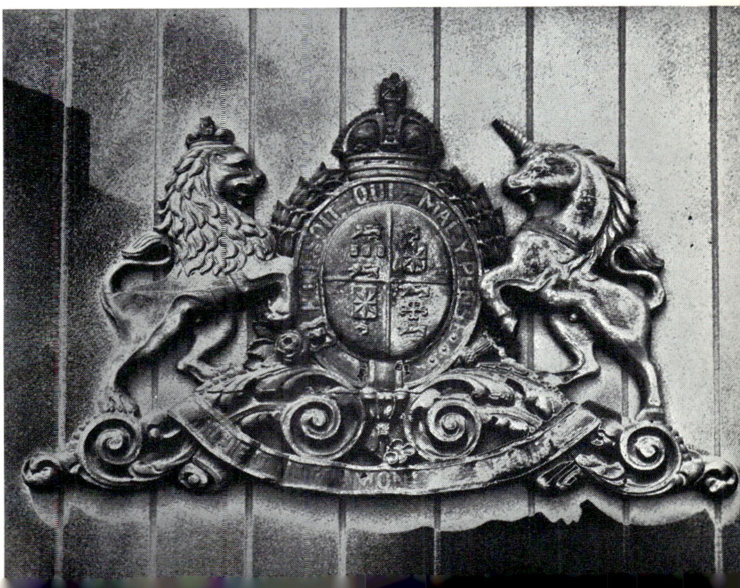

5-2D

5-2G

5-2F

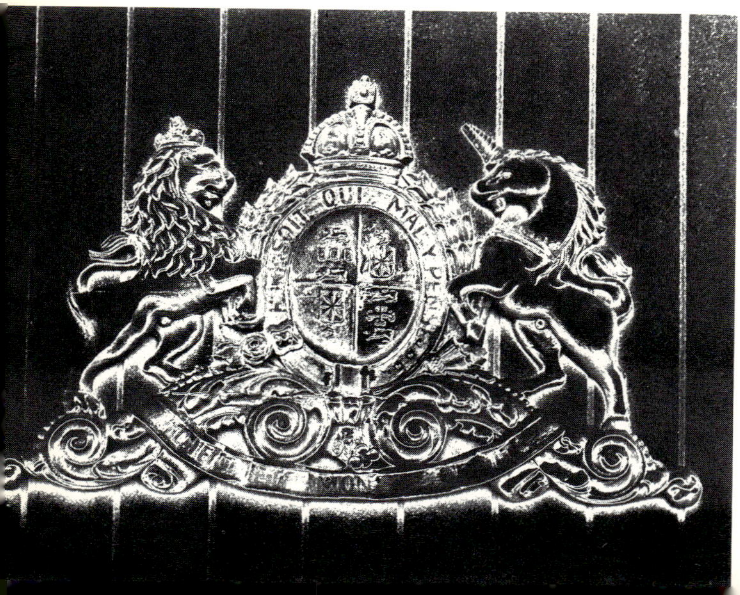

5-2E

negative was of normal density, but somewhat flat. We have given print Fig. 5-2G a touch of gentle solarization by treating it like Fig. 5-2D but reducing the second exposure to 5 seconds with a 15-watt bulb.

The third principal variable involved in the control of the Sabattier effect is the second exposure; the possible variations in the second exposure are intensity, color, and duration. In black-and-white photography, we found no particular effects associated with the color of the exposing light, aside from the fact that the emulsion varies in sensitivity and requires more or less exposure depending on the particular wavelengths in the light. The intensity and time of the exposure, however, are of considerable importance, separately and in concert. If you wish the solarized area to go completely black, you must use a short exposure to a high-intensity bulb. When the developer is working well, we find that at a distance of five feet a 40-watt bulb is the minimum that will give good blacks, and if you want to insure getting them it is safer to work with a bulb of 75 or 100 watts. A developer that produces good blacks with less exposure is working too fast, and is not aged sufficently to give the highest quality solarizations. Using a 25-watt bulb at five feet, we found that an unexposed piece of paper would not go black even after a five-minute development with the light on, provided it had been in the developer at least 30 seconds before the exposure. When the solarizing light is turned on, the emulsion begins developing at once because it is saturated with developer. But in the course of development, the rest of the emulsion becomes progressively desensitized, so that by the end of a one-minute exposure the material is practically insensitive to light. To illustrate some of the effects of the second exposure on a solarized print, we have used our cover negative, which has an opaque background. Each of the prints in Figs. 5-3A to E was given a 10-second printing exposure, a 45-second first development and a 75-second final development after solarizing. Only the solarizing exposure was different in each case.

Fig. 5-3A was solarized by a 15-second exposure to a 25-watt light bulb; Fig. 5-3B was exposed continuously to the same light during the entire second development. Comparison of these illustrations demonstrates that the second exposure inhibits the development of the first exposure.

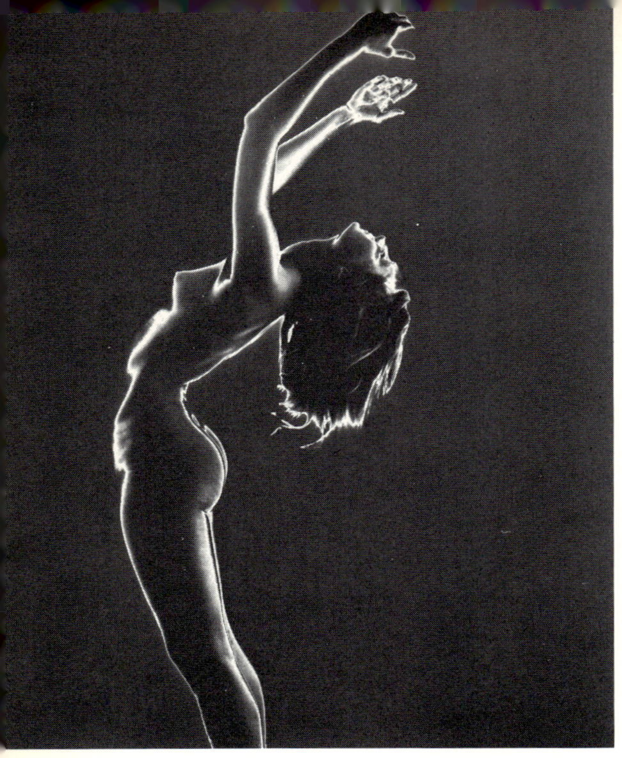

5-3A

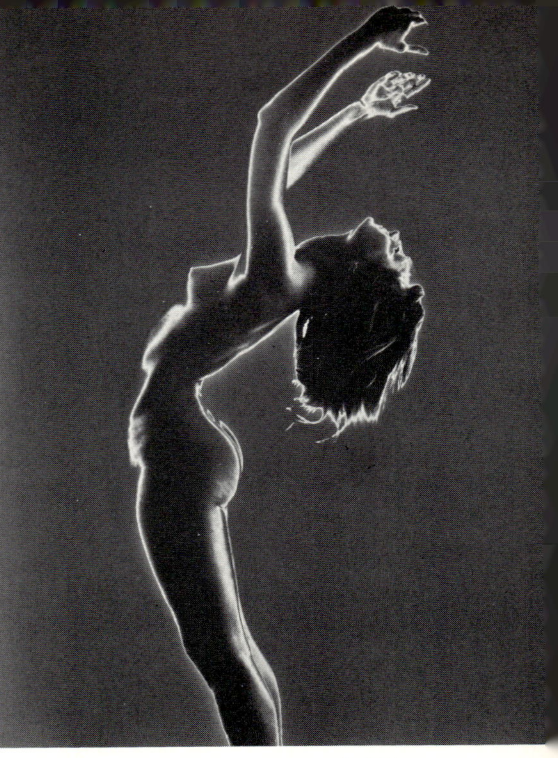

5-3B

Since the longer exposure gives the lesser density, this effect could be mistaken for *true* solarization, which has this characteristic. But this case is different because the second exposure is given while the print is immersed in the developer, and the two similar effects may well have different causes. Fig. 5-3C was treated to a second exposure consisting of an electronic flash discharge two feet above the tray. The same flash was used in the bounce mode for Fig. 5-3D, resulting in a much weaker exposure. The white speckles in Fig. 5-3D are the result of the floating debris in the old developer having been frozen by the very rapid (1 1000 second) flash. During the usual exposure of several seconds, the tray is rocked to keep the particles in motion so that their shadows do not register on the print.

One of the popular explanations of the Sabattier effect is that the image developed during the first exposure shades the emulsion beneath it and prevents it from receiving much of the second exposure. That is, the first image is printed by the second exposure, thus causing the reversal. This may contribute to, but certainly does not account for, the entire effect, as has been demonstrated by a number of studies. In Fig. 5-3E we show a solarized image that was produced when

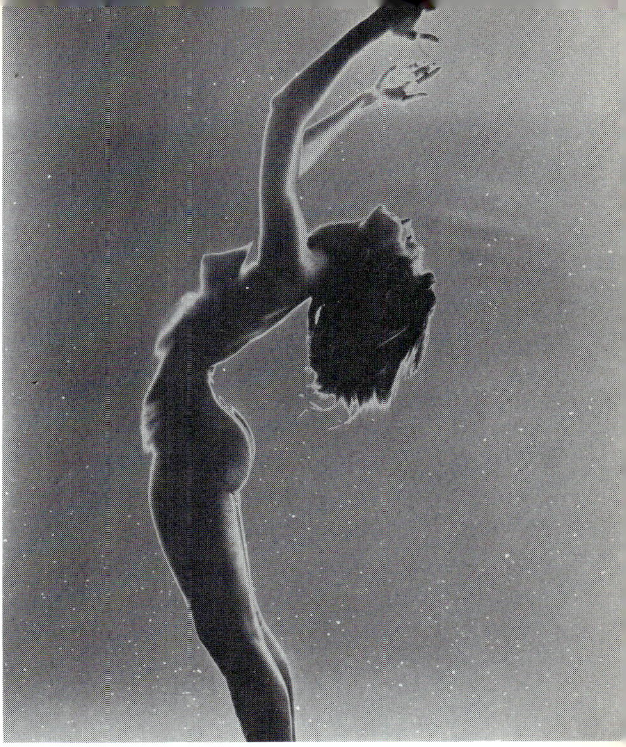

5-3D

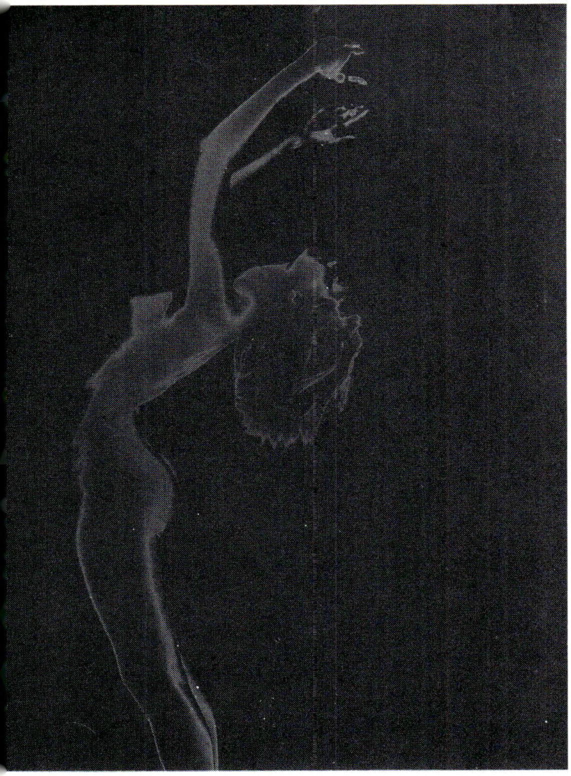

5-3C

5-3E

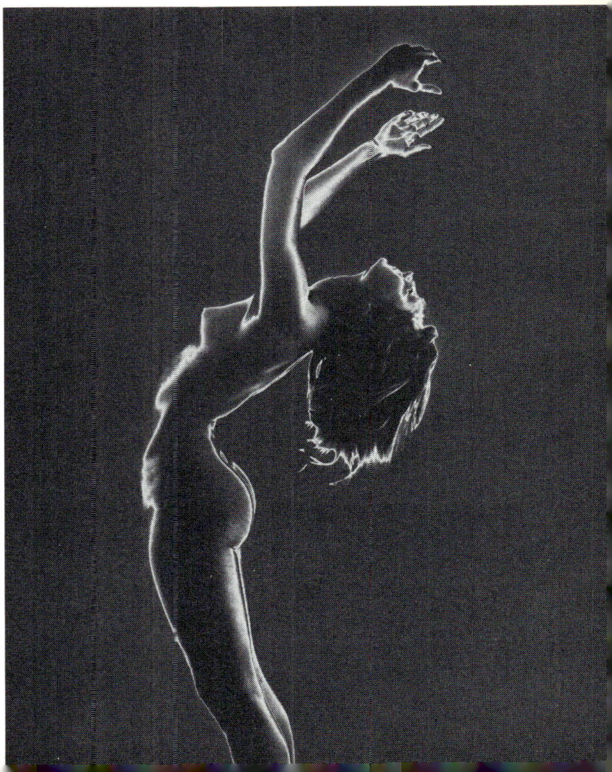

the second exposure was given from the back of the print. This picture was treated just like Fig. 5-3A except that the second exposure was 60 seconds long with a 40-watt light bulb at a distance of three feet above the print, which was face down in the black tray. The only visible difference between Fig. 5-3E and Fig. 5-3A is that in 5-3E a slight mottle appears in the background due to the paper fibers.

The same negative can be used to illustrate another interesting, and possibly useful, feature of the Sabattier effect. This involves our gentle solarization process, in which the second exposure is given with a 15-watt bulb, and the image is enhanced without being reversed. Fig. 5-4A is an unretouched straight print made on No. 2 contrast paper from this negative. Notice that the background is completely clear white and that the highlight on the model's belly almost blends with it. As a matter of fact, the negative originally was not opaque in the background, but transmitted enough light to cause fog on a normal print. To eliminate this, a Kodak opaquing compound was applied rather carelessly to the entire negative except for a small area around the figure. Badly done though it was, the treatment was effective, and for normal printing there was no "show through." But now look at Fig. 5-4B where the same negative has been printed by the gentle solarization technique on No. 6 paper. Not only has the belly highlight been recovered from the background, but the poor job of opaquing has been disclosed. To make this print, the first exposure was about half of normal so that the full development showed a pale image. The 15-watt solarizing light was turned on for five seconds only, and then the print was allowed to develop for another minute to reach this stage. It seems that the second exposure is very effective in bringing up the invisible, sub-latent image that would ordinarily not develop at all, and since the rest of the image has already developed before the second exposure is given, it does not result in a generally fogged print. We know of no other printing procedure that is so effective in extracting information from a negative, and we believe this technique may be the most important outcome of our work with the Sabattier effect.

The fourth principal variable, the second development, does not concern us greatly. As in normal printing tech-

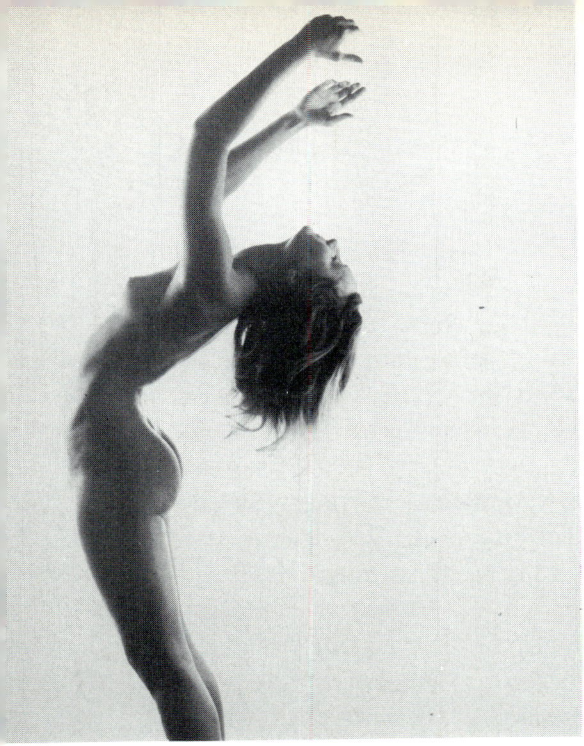

5-4A

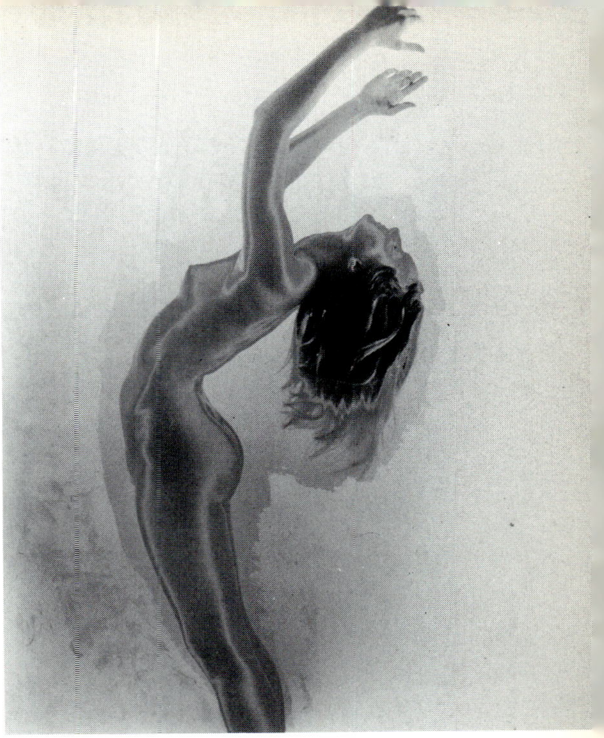

5-4B

nique, good print quality is seldom achieved by stopping development before it is completed. If the exposures have been handled properly, there is no need to snatch the print from the tray to prevent overdevelopment, and the final development time should always be long enough to bring up all of the image that has been exposed. This usually means at least one minute, and sometimes, in the case of our ''old brown'' mixture, as much as three minutes. With a correctly exposed print there is always an easily recognized time at which image quality has ceased to improve but fogging has not yet begun. That is the time to remove the print from the developer.

The type of paper you choose has quite a bit to do with the kind of solarizations you get. We have tried several of the popular brands of enlarging papers, a few contact papers, and some special-purpose types such as reflex copy, panchromatic, and one with a very slow emulsion designed for line copy. In general, the higher the contrast grade of paper, the better the contrast in the solarized print, and the better the edge effects. Low contrast papers, particularly portrait papers, have a tendency to produce a negative, but can be made to give interesting results if the proper exposure conditions are found. They almost always require reprinting in order to achieve acceptable contrast. Variable contrast papers seem to behave similarly, and there is some improvement when the filters are used to increase the contrast. All the papers showed the Sabattier effect to some extent, including the very slow one, which had to be exposed under a photoflood bulb and similarly solarized. Each type of material seems to have its own peculiar characteristics and possibilities for individual effects, but with none of them was the process so delightfully easy and satisfying as with the very high contrast bromide enlarging papers. While papers of lower contrast are capable of producing some very delicate effects, they are less versatile and require more critical conditions of exposure.

From the beginning of our involvement with the Sabattier effect, the necessity for "old brown" developer has been most intriguing. Why does it give better results than fresh developer? Is it possible to add something to or leave something out of the formula for fresh developer to make it work just as well? Does this throw any light on the mechanism of the Sabattier effect itself? We cannot promise any definite answers here, but only an account of some of our observations and thoughts on the subject.

One of the most obvious things about "old brown" is that its action is considerably slower than that of fresh developer. Perhaps it would suffice, then, to slow down the action of the fresh solution by diluting it more than usual. We tested progressively more dilute solutions, going from 1:1 to 20:1, and found that the solarizing properties improved slightly, but the image quality decreased until, at high dilution, it was impossible to get a real black at all.

The next experiment was to deliberately oxidize the developer rather than age it by use. Adding increasing amounts of 10 percent hydrogen peroxide served only to oxidize the developer without appreciably improving its solarizing ability. It also quickly killed the activity of the solution, and turned it dark opaque brown—not the transparent coffee color of our old faithful. Using Clorox (sodium hypochlorite), a household bleach, as an oxidizing agent, we were able to produce a few adequately solarized prints, but the solution was not stable. Once the oxidation had begun, it would continue progressively until the developer was a dead opaque brown, just as before. There was a brief stage during the process when it behaved as a solarizing bath, but we were unable to halt it at that point, and each time, in less than 20 minutes, it had expired. In the course of our experiments with Clorox, we had added progressively larger and larger quantities to the developer. For a grand finale, we placed a print in a tray of strong solution of bleach. A few seconds later, we hurriedly dumped the tray into the sink and dashed out to escape the chlorine fumes that filled the darkroom. So we offer a slight warning: A strong solution of bleach will attack the gelatin, cause frilling on the edges of the print, and give off noxious fumes.

Aerial oxidation was another possibility, and a simple one to test. An opened can of Dektol was allowed to oxidize in powder form, and then used to mix the developer. Aside from being dark brown, the solution behaved very much like the usual fresh Dektol, developing prints adequately, but solarizing poorly. In another experiment, a tray of fresh Dektol, diluted 2:1, was left side by side with a tray of regular "old brown" that had been well used for solarizing. After ten days, the two were poured into glass beakers and compared. "Old brown" had changed very little. It was slightly darker than before, but still transparent. The unused developer, however, was a dark, totally opaque brown. The two solutions were returned to their trays and tested for solarizing properties. Neither of them would develop a normally exposed print within any reasonable time. No image was visible after five minutes in either solution. The two mixtures gradually began to regain their activity as fresh stock Dektol was added ounce by ounce. There had originally been 24 ounces of solution in each tray. When six ounces of fresh stock had been added to each, "old brown" was able to solarize prints adequately, although not as well as before the experiment had begun. The unused developer, however, solarized the same way as freshly mixed solution and yielded dark foggy prints. Apparently, the magic ingredient of "old brown" is not simple oxidation.

Among the various chemicals that we added to fresh developer in trying to produce the effect of "old brown," the most reasonably promising was potassium bromide. This chemical is widely known as a restrainer that is used in most developers to prevent fog and thus increase contrast. Since the bromide concentration of a developer normally increases with its use, it seems possible that the efficacy of "old brown" could be due to its high bromide content. In a tray containing eight ounces of fresh developer stock and sixteen ounces of water, we added potassium bromide in steps of two and one-half grams. The first effect was a slowing down of the activity of the developer. At five grams, a noticeable improvement in solarizing ability was evident. At ten grams, the solarizing action was fairly good, though not quite up to the standard of "old brown," but the developer was very slow and the quality of the image was beginning to suffer. Beyond this point the action of the developer became very slow, and there seemed to be no further improvement in solarizing ability. Our conclusion is that this is a useful emergency technique. If you want to solarize immediately, but have no used developer, just add about five grams of potassium bromide to each pint of your working developer. The results will not be the very best, but they will be very much better than you can get with fresh developer alone.

Unable to imitate the quality of "old brown" by the use of an additive, we decided to try the opposite approach. Perhaps its secret was in the depletion of some constituent during use. Our next series of experiments, then, was to mix the developer according to the formula for D-72, but leave out one constituent each time. Five solutions, each with one ingredient missing, were made and tested for solarizing ability. The results were completely unpromising, and largely unpalatable as well.

The next step was to make up the developer chemical by chemical, testing its ability after each addition. Again, the results were negative, although there was one interesting phenomenon that was new to us. Using the solution containing only hydroquinone, metol, and sodium sulfite (no sodium carbonate or potassium bromide), the developing action was very slow, since there was no accelerator. Accordingly, after waiting three minutes for the image to reach an adequate density, we turned on the light and left it on. Very gradually, over the course of the next four minutes, the reversal began to occur, but the second image was brownish-purple, like the image on a printing out paper. When we placed the print in the fixing bath, we were amazed to see the image re-reverse—that is, the solarized image due to the second exposure just bleached right out, leaving the print with only the first exposure image. This result was so unexpected and so fascinating to watch that we repeated it several times before continuing with the rest of the program. The effect disappeared entirely after the addition of the sodium carbonate, and the remainder of the experiments produced routine results.

We explain the fading of the solarized image by noting that the slow growth of the image resulted in a very fine-grained deposit of silver consisting of incompletely developed specks, which not only accounts for the brownish color, but also for the high solubility in the fixing bath. Just why the solarized image should be different in this respect from the original image is another matter—one for which we have, as yet, no explanation.

Since none of our experiments turned up a good substitute for "old brown," we continue to use it for nearly all our work, and have found it very reliable and no real hardship to maintain near optimum effectiveness. When it becomes too tired to produce a good print in a reasonable time, say in three minutes, we add a few ounces of fresh stock to bring it back to life. When it gets too dark or too deep for the light to penetrate it, we remove half of it and add one-half that amount of working-strength fresh developer. The discarded "old brown" is not thrown away, but kept in a large jug, where it continues to get darker and acquires a sickeningly sweetish odor. We find that a satisfactory solarizing mixture can be made by combining one part of this old stuff with one part of fresh developer stock and four parts of water. (At this dilution, the odor is not objectionable.) We have observed that the solarizing ability is not directly related to the dark color of the developer, but it does require that the solution be thoroughly used to a point far beyond the manufacturer's recommendation. A tray of developer containing eight ounces of stock and sixteen ounces of water will begin to solarize well after it has developed about forty 8″ x 10″ prints, although Kodak would probably have you throw it out after thirty. The exhaustion of a developer is partly due to the depletion of the developing agents, but a more important factor is the increasing concentration of the reaction products, which inhibit the developing action. In addition to the oxidized metol and hydroquinone, the used developer contains silver, bromide, and hydrogen ions, and has become slightly more acidic than before. Unused developer that has been allowed to oxidize contains only hydroxyl ions in addition to the oxidized developing agents, and is thus slightly more alkaline. Since the used developer solarizes well, while the unused does not, the difference must be due to the silver, the bromide ions, or the acidity, or to the combination of these. There is no evidence that the acidity affects the solarizing ability other than slowing the speed of the development, and we have already seen that the addition of bromide ions in quantity improves this ability insufficiently, and at the expense of other virtues. We are therefore led to suspect that the magic ingredient of "old brown" may well be just the dissolved silver that accumulates in the developer when it is much used.

Scientific studies of the Sabattier effect have most often been carried out under conditions carefully restricted so that the effect would occur in its purest and simplest form, thus making the job of interpretation much more straightforward and less subject to error. Less attention has been paid to the Sabattier effect under working conditions such as we have here because it is thought that the complex nature of the reactions makes it unlikely that such study would lead to an understanding of the basic photochemical mechanisms. It seems quite clear that a number of different causes are involved in producing the varied effects that we have lumped under the name of Sabattier, and the proper scientific approach is to isolate these causes one at a time and study the effect of each one alone. Our aim is not a scientific knowledge of the chemistry and physics of the photographic process, but to gain some practical idea of the possible causes of the Sabattier effect in order to use it more effectively.

It may be helpful to summarize the observations that we wish to discuss:

1. The first exposure partially desensitizes the emulsion to the second exposure by an amount somehow related to the density of the developed image.

2. There is usually a border effect consisting of a band of lower density between the image developed after the first exposure and that developed after the second exposure. When this band is very sharp and narrow, we have called it the Mackie line.

3. If the second exposure is given before the first exposure is completely developed, the final density of the first exposed area may be less than it would have been in the absence of a second exposure.

4. It is possible by means of a very weak second exposure to enhance the sub-latent image to the point of developability without generally fogging the print or causing reversal.

5. The above effects are emphasized by the use of a well-aged developer that has been used beyond the point of normal discard.

The desensitization mentioned in the first observation is the basic feature of the Sabattier effect. Different investigators have shown that it can have several possible causes. Among the most likely of these are the following:

a. The first exposure uses up the most sensitive silver halide grains and thus leaves the image area in a less sensitive condition.

b. The first exposure image shields the underlying emulsion from the second exposure light.

c. Bromide ions released during development of the image act locally to restrain further development in the same image.

d. The unexposed silver halide grains are poisoned by the silver produced in the development of neighboring grains.

e. Developer oxidation products build up locally in the developed regions and restrain further development.

It is certainly possible that any or all of these causes are at work in our case, since we have done nothing to eliminate any of them. We can only guess as to which of them is the dominant one, but the use of "old brown" would have no influence on causes *a* and *b* but might be expected to increase any effects due to *c, d,* and *e.*

Remember that this developer is well on the way to exhaustion, and not only is it high in oxidation products, bromide ions, and silver, but it is low in active developer molecules. After the first development, the denser image areas are saturated with exhausted developer while the lighter or unexposed areas are saturated with active developer. When the second exposure begins, the active developer goes to work immediately, while the exhausted developer in the first image must be replaced by diffusion before any development can take place there. Thus the new, solarized image has a head start on the old image in the development process, as has been observed. This can not be the whole story, however, because we have often seen a print reach a certain stage of development at which it seems to become almost insensitive to further exposure. This indicates a general desensitization of the entire emulsion, in addition to the local image desensitization. We assume, then, that nearly all of the silver halide grains have been exposed, but have in some way been prevented from developing. We might guess that the preventing agent is metallic silver, which has infected each of the grains of the emulsion. Just how silver can be a desensitizing agent is not really known, but it has been suggested as such in the explanation of the Sabattier effect given by Stevens and Norrish. (See Bibliography.) They suggest that an atom of silver deposited on an unexposed silver halide crystal competes with the sensitivity speck for the collection of silver atoms released by the action of light on the crystal. In order for a silver halide crystal to be developable, it must contain a latent-image speck consisting of some minimum number of silver atoms, say about four. These atoms are produced one at a time, each time the crystal absorbs a photon of light, the first formed acting as a center to trap the second and so on, until the speck has grown large enough to initiate the development of the crystal.

Suppose, however, that some other process deposits an atom of silver at another or perhaps at several other points in the crystal. These atoms will act as competitors to the light-produced speck, and reduce its chances of acquiring enough additional atoms to become a latent-image speck. For instance, a crystal that contains six silver atoms in two separate specks of three atoms each will not develop, whereas a crystal that contains five atoms in one speck will.

Lest you wonder why the atoms don't all combine to form one large speck, we should explain that the trapping of atoms by a sensitivity speck is thought to be at least a two-step process. The silver bromide crystal is a lattice of positive silver ions (silver atoms with one electron missing) and negative bromide ions (bromine atoms with an extra electron). The ions are held in their places by the electrostatic forces between their charges. The action of the light is to free the extra electron from the bromide ion, which then becomes a neutral atom and may diffuse out of the crystal into the gelatin. The freed electron wanders through the crystal until it is trapped in some imperfection or sensitivity speck in the crystal lattice, giving that spot an extra negative charge. There it attracts to itself one of the silver ions that has been freed from the lattice by thermal motion. The silver ion combines with the electron to form a neutral atom of silver, which then acts in turn to trap the next electron and silver ion.

It may happen that, due to thermal energy, a trapped electron may escape again and possibly recombine with a bromine atom, so the building of a latent-image speck is a statistical process depending upon the balance of competing events. Anything which lowers the chances of a single speck of growing to developable size will act as a desensitizer. Thus, excess bromine, which increases the possibility that the free electron will recombine with a bromine atom, or silver atoms, which act as competing nuclei, might retard development of exposed grains.

The above description of the mechanism of latent-image formation is very much simplified, and bears about the same relation to the formal theory as does a quick sketch on the tablecloth to an architect's plans. The purpose of our sketch has been to illustrate the kind of problem we are dealing with in the Sabattier effect. The fact is that, in spite of their careful efforts to eliminate all other possible causes, Stevens and Norrish were not able to prove that the desensitizing agent was metallic silver; it has not even been demonstrated that metallic silver can act as the desensitizing agent. In our case, we suspect that the mechanism described above is not responsible for the results we observe, for it seems to imply that the desensitization would be the same for fresh developer as for old. We have noticed that, after reaching the point where the print in the old developer has effectively stopped developing even though it is continually being exposed, if we transfer the print to fresh developer it will resume development and acquire a general fog. This indicates to us that the explanation we seek must lie at a different level, and is probably not to be found in the secondary photolytic process within the crystal, but more likely in the ability of the developer to attack the exposed crystal and reduce the silver bromide.

One of the interesting features of the Sabattier effect

with "old brown" developer is the difference between the image developed after the first exposure and the one developed after the second exposure. We have already mentioned that the solarized image may be a different color and is more easily bleached. We now note that it is more subject to scratching by print tongs, and is more susceptible to fingerprints. This difference must be due to the fact that the first development acts on pre-exposed silver bromide crystals in which all photolytic processes are completed and that are not encumbered by reaction products of early development. The second exposure, on the other hand, takes place while the crystals are already surrounded by active developer molecules as well as debris from the previous development. Each latent-image speck, as it appears, is subject to immediate action by the developer, mitigated by any influence exerted by the previous reaction products. It seems possible that these different conditions could favor a different type of development, or, rather, shift the balance between the two types of development that are taking place. The most usual type is called direct or chemical development, and takes place at the interface between the silver latent-image speck and the silver halide crystal. Direct development does not require any silver or silver ions in the solution, and is probably the major process responsible for development under practical conditions. When the developer solution also contains free silver ions and a compound such as sulfite, which will form a complex with the silver ions, we have the conditions for physical development, in which metallic silver is deposited from the solution onto the latent-image nuclei. It is characteristic of the image produced by physical development that it is very fine-grained and easily dissolved. Moreover, there is a tendency for the silver to plate out on the print surface in such a way that it is susceptible to abrasion damage. It also plates out on the developing trays, a phenomenon we have often noticed with "old brown." Physical development is favored by weakly acid conditions, requires only a weak reducing agent, and needs several times the exposure required by chemical development.

The above considerations lead us to believe that, while the first development may be primarily of the chemical type, the second development is predominantly physical. This speculation is based on purely qualitative observations, and we have no way of knowing what percentages of each type of development are involved in each stage. Nor do these observations help us with the still unanswered problem of the nature of the desensitizing agent that must be at the root of the Sabattier effect. Assuming that there is some agent that desensitizes the emulsion in proportion to the developed image, it is fairly easy to explain the commonly observed border effect in solarized images by reference to the sensitometric curve in which the developed density is plotted against the logarithm of the exposure. If we have a full tone scale in the print due to the first exposure, some point on the print will correspond in density to each of the points on the curve, which looks like curve "A" in Fig. 5-5.

If the desensitization is assumed to be proportional to the developed density, the uniform second exposure will produce a density that, when plotted on the same graph against the logarithm of the first exposure, will give a curve that is the mirror image of the first. This is shown in Fig. 5-5 as curve "B," while curve "A" is the original D–log E curve for the first exposure. We are assuming here that the second exposure has not affected the course of development of the first. The density due to the first and second exposures together is represented on the same graph by curve "C," which is the simple sum of the other two curves. This shows

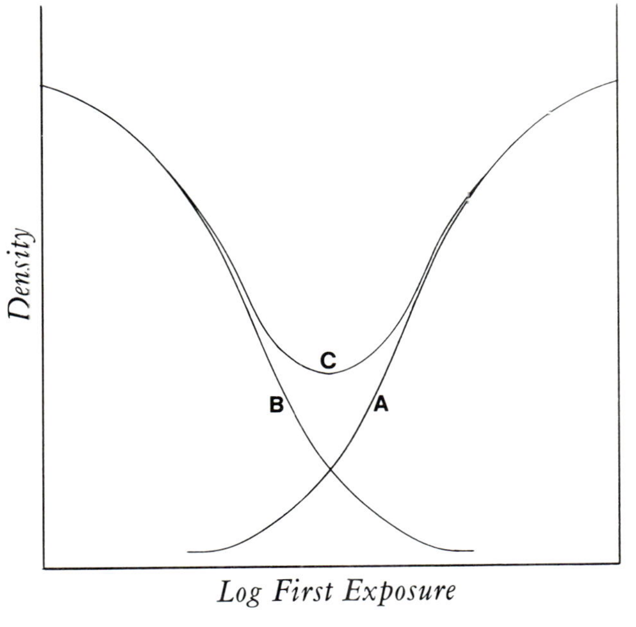

5-5

a minimum density where the two curves cross, indicating the border of lighter density that is observed on the print between the two images. That part of the image whose densities lie on the left half of the curve "C" will be completely reversed, since the lower exposures will have the higher density.

The explanation above was given by Stevens and Norrish in 1937. It not only demonstrates the basic phenomenon of the Sabattier effect, but also shows the dependence of the effect on the characteristic curve of the emulsion. Suppose, for example, that we choose a printing paper whose D–log E curve looks like "A" in Fig. 5-6. When we include the second exposure curve and draw the combined result, we get something like curve "C" in Fig. 5-6. It is evident that this emulsion would solarize very poorly because the sum of the two densities at the crossing point is almost equal to the maximum density at the extremes of exposure, and the contrast would thus be close to zero.

Stevens and Norrish extended their explanation to include the narrow Mackie line that occurs when the image is that of a sharp edge. They attributed it to the Sabattier reversal of the image formed by scattered light in the emulsion, claiming that the abrupt edge was in reality a steep density gradient that, upon solarizing, exhibited the

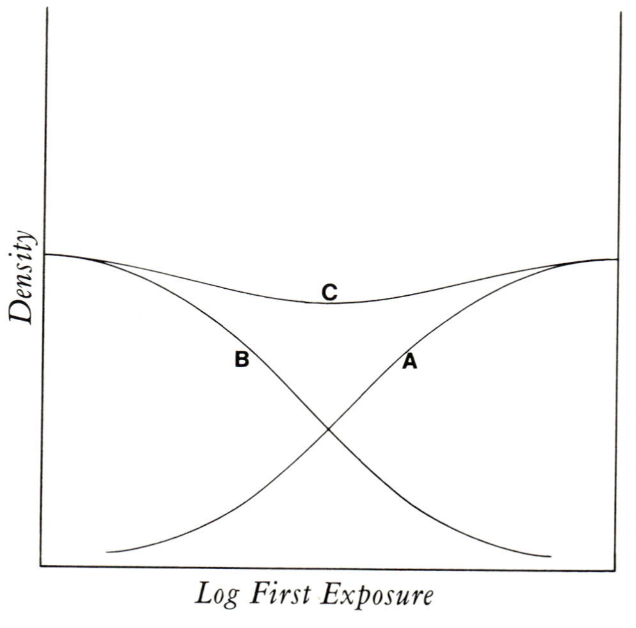

5-6

same minimum density as described above, but in a very narrow region. All of their work was done with emulsions coated on film, and their procedure included removing the film from the developer and washing it before the second exposure.

In our own experience, it seems unlikely that there is anything like an exact proportionality between the amount of desensitization and the density of the first developed image. Rather, it seems to us that the desensitization is related to the developed density in a more complicated manner and may also depend somewhat upon the first exposure itself. Using a series of constant exposure increments while the paper was immersed in the developer, we frequently found two obvious minima of density rather than one. One of these test strips is shown in Fig. 5-7. The interpretation of this is complicated by the fact that it can be considered as a multiple solarization, since a ten-second development time was allowed between exposures. It demonstrates very clearly, however, that the original exposure development is inhibited by the later exposure and development. This is not to say that the latent image formed by the first exposure is destroyed by the second. It indicates that the development of the first exposure is interfered with by the reaction products from the development of the second exposure, which is taking place simultaneously. This is apparently the explanation of summary observation No. 3 that appeared earlier in this chapter.

The enhancement of the sub-latent image, as mentioned in observation No. 4, is just what one might expect to happen when additional exposure is given. By providing a very weak second exposure, below the threshold of the completely unexposed area, the sub-latent image in the highlights is allowed to develop, while the Sabattier desensitization holds back the development of the shadows. One interesting consequence of this is the fine black line that appears in place of the usual white Mackie line. (See Fig. 3-26B.) This is not the same line, for careful examination with a magnifying glass will show that it lies next to but outside of the white line. It is not explained by the analysis of Stevens and Norrish, but is apparently an extreme example of what is called in proper parlance the "border" effect. It is this border that, when printed, gives the white line that

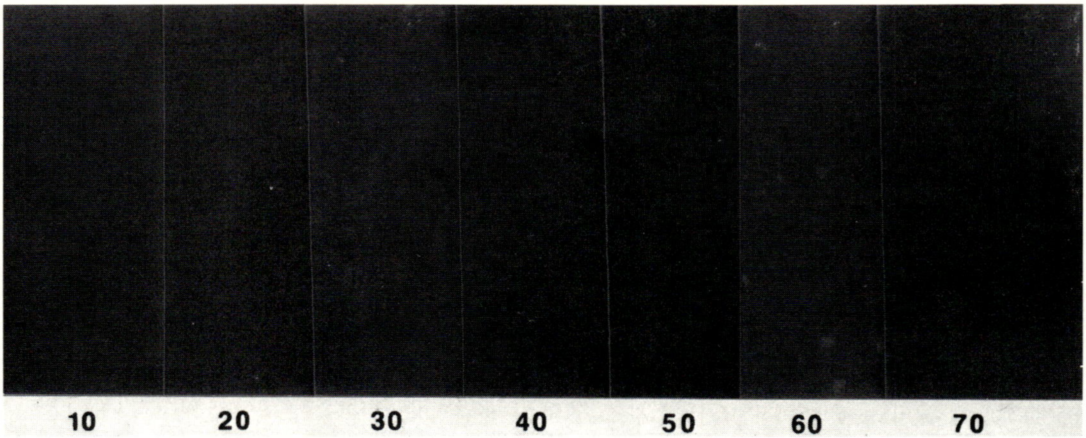

Figure 5-7. An unexposed sheet of printing paper was immersed in a tray of used developer for one minute before the series of exposures was begun. An opaque plastic board was used to cover part of the paper, while the rest was exposed to the light of a 25-watt bulb five feet above the tray. The exposure sequence was: ten seconds on, ten seconds off, ten seconds on, ten seconds off, and so on. During each ten-second dark period, the plastic board was moved to uncover about two centimeters more of the paper. The resulting total exposures of the strips, reading from left to right, was 10, 20, 30, 40, 50, 60, and 70 seconds, respectively. The development time after the final exposure was one minute.

---

is called the Mackie line in technical language. It is thought to be due to the diffusion of fresh developer from an unexposed region into an adjacent, lightly exposed one. These conditions are fulfilled here if we can equate the highlight area with the unexposed region. The highlight area has been exposed, of course, but because of the Sabattier desensitization, no development has taken place. This means that there are unused developer molecules in the highlight that can infuse into the newly exposed region, which has not yet been affected by the desensitization. If this is so, then we must conclude that the desensitization agent does not diffuse as well. In any case, in view of the plentiful supply of active molecules in the developer, the effect of the supposed diffusion across the border seems excessive. Whether or not the above is the correct explanation of the fine black line, the effect promises to be a very useful one for separating an image from its background and increasing the apparent contrast without blocking up the shadows. The exposures required are quite critical, and must be determined by test for each negative. We have found that a first exposure about two-thirds normal is a good starting point. The print is developed two-thirds full time and the second exposure is made for about three seconds to a 15-watt bulb at a distance of five feet. Final development should be approximately as long as the first. A shorter first exposure coupled with a slightly greater second exposure will result in a grey, partly reversed image having the appearance of metallic silver. The male torso in Fig. 5-8 (next page) has been given this treatment. There still remains a trace of the black line along the inside of the upper right arm.

As we conclude our task of setting down these notes, we are very much aware of how many questions have been raised and how few are the answers we have been able to give. In spite of the tests and measurements, in spite of the theory and the graphs, the making of photographs, solarized or not, remains an art and a craft—not a science. Nearly three years have passed since we rescued that first crazy print from the trash can. We can now go into the darkroom and reproduce that effect and many others at will. The prints we have solarized number well into the thousands, and we are completely familiar with all the strange things that can happen in Sabattier land. But familiarity is not understanding, and we still can't tell you, for sure, why they happen that way. No one knows, as yet, and perhaps that's a good thing. It will be a sad day when all questions are answered and all mysteries are solved, for then you will no longer be able to see that print developing before your eyes and believe that it is pure magic, and that's really what it is.

5-8

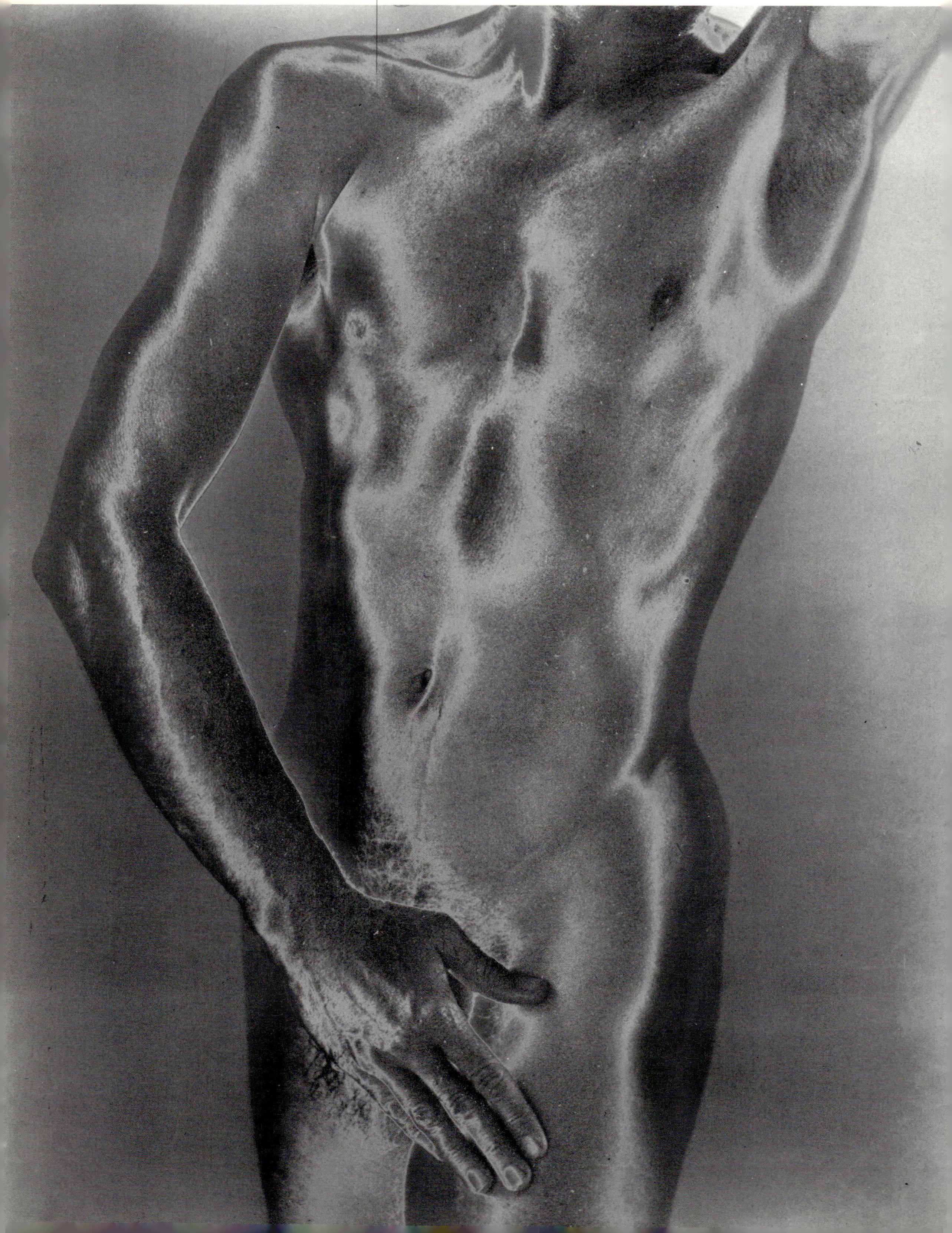

A Gallery of Solarized Prints

The following notes apply to some of the pictures in this section:

(Page 132, Gallery Print 1)
Tone separation. The diapositive was used for the dark areas and a soft positive for the grey areas.

(Page 133, Gallery Print 2)
The model in this picture was actually doing a backbend, supported on her left arm in front of a white wall. When the original print was solarized, the white wall became black, but the figure reversed only slightly because it was already well-developed when the second exposure was made. This is a straight print from the solarized original and shows the characteristic reversed Mackie line. There was an electrical outlet in the wall just before the model's face. It was held back during the enlarger exposure, and completely obliterated by the solarization. Had it persisted after solarization, it could have been eliminated by opaquing it on the original before making this print.

(Page 134, Gallery Print 3)
This salt factory was taken in daylight. Previously the smoke was lighter at the top but now, reversed, it is as if lit from below by the setting sun.

(Page 135, Gallery Print 5)

After the print is in water for 24 hours, the dryer apron artistically removes some of the emulsion.

(Page 136, Gallery Print 6)

Low contrast solarization on portrait proof paper. This is the negative version.

(Page 137, Gallery Print 7)

Originally photographed against a white background, this Canadian carved wood duck was texturally enhanced by the addition of a fine etching screen before solarizing.

(Page 138, Gallery Print 8)

Poor registration in the printing and natural decay of the house together create this woeful scene.

(Page 140, Gallery Print 10)

Here is an example of natural 35mm zoom lens grain, much improved by solarizing.

(Page 140, Gallery Print 11)

A short development after the solarizing exposure will give a brown tone in the redeveloped areas. These reproduce as grey.

(Page 142, Gallery Print 13)

This grim scene is of the mudflat sculptures on the San Francisco Bay.

(Page 143, Gallery Print 14)

This is a very thin negative with very thick dirt. Instead of cleaning it, we solarized it.

(Page 144, Gallery Print 15)

Two different negatives of the same cat were sandwiched together in the enlarger to make this picture. The cat was strongly backlighted so that no detail showed in the face, and the background was dark. This is a second generation print from the solarized original, and looks just like it except for some improvement in contrast. To get the grey tones on the cat's back, No. 4 paper was used for all three prints.

(Page 145, Gallery Print 16)

An abstract was made by photographing a figurine through

a glass block. The negative was sandwiched with that of the model and solarized on No. 6 paper. This first generation print from the original has been bleached in the surrounding areas to remove unwanted grey tone from the white area.

(Page 146, Gallery Print 17)

Night photography without flash. Take a daylight picture and solarize.

(Page 149, Gallery Print 20)

A negative sandwich of a face and an abstract was double printed with a fireworks explosion, and the whole thing was solarized and printed again.

(Page 151, Gallery Print 22)

In the final print the solarized area is light brown due to the removal of the print from the developer before solarizing was complete. In black-and-white reproduction this area is rendered as a grey.

(Page 153, Gallery Print 24)

Two prints were taken from a solarized original, one somewhat darker than the other. These two were placed in register and printed to give a rather light image in apparent bas relief. This light print was then printed again to produce this final result. The harsh, stone-like texture is due to superposition of the grain of the two registered prints. Of course, when printing through two pieces of paper, it is impossible to have all three emulsions in contact at once, so there is necessarily some light scatter between the pair not touching.

(Page 154, Gallery Print 26)

This is a result of combining a negative of some mudflat sculptures with one of a face painted on a construction fence. Solarized and reprinted, the whole forms a weird abstraction, bearing little resemblance to the originals.

(Page 156, Gallery Print 28)

This is a solarized composite of four negatives.

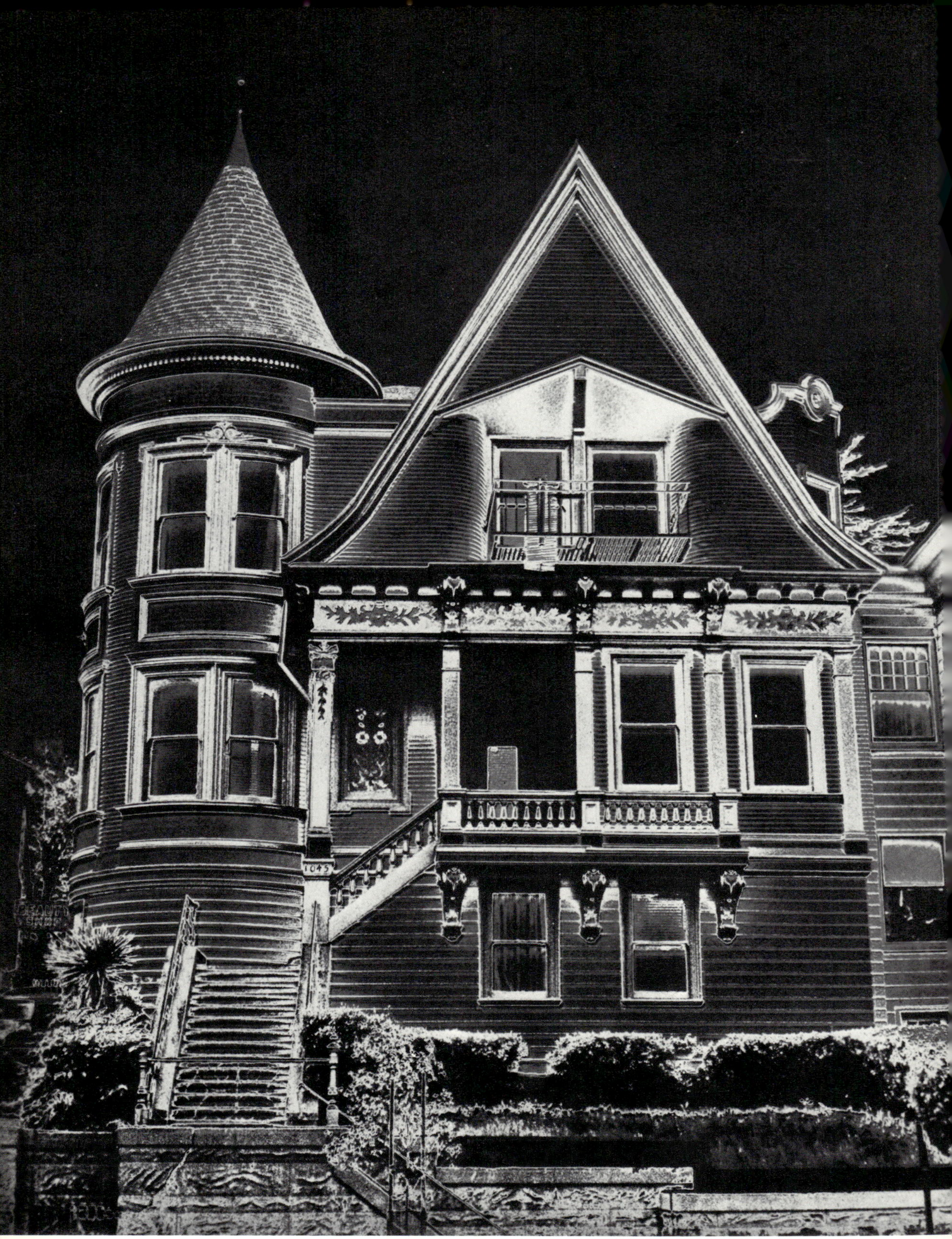

GP-1

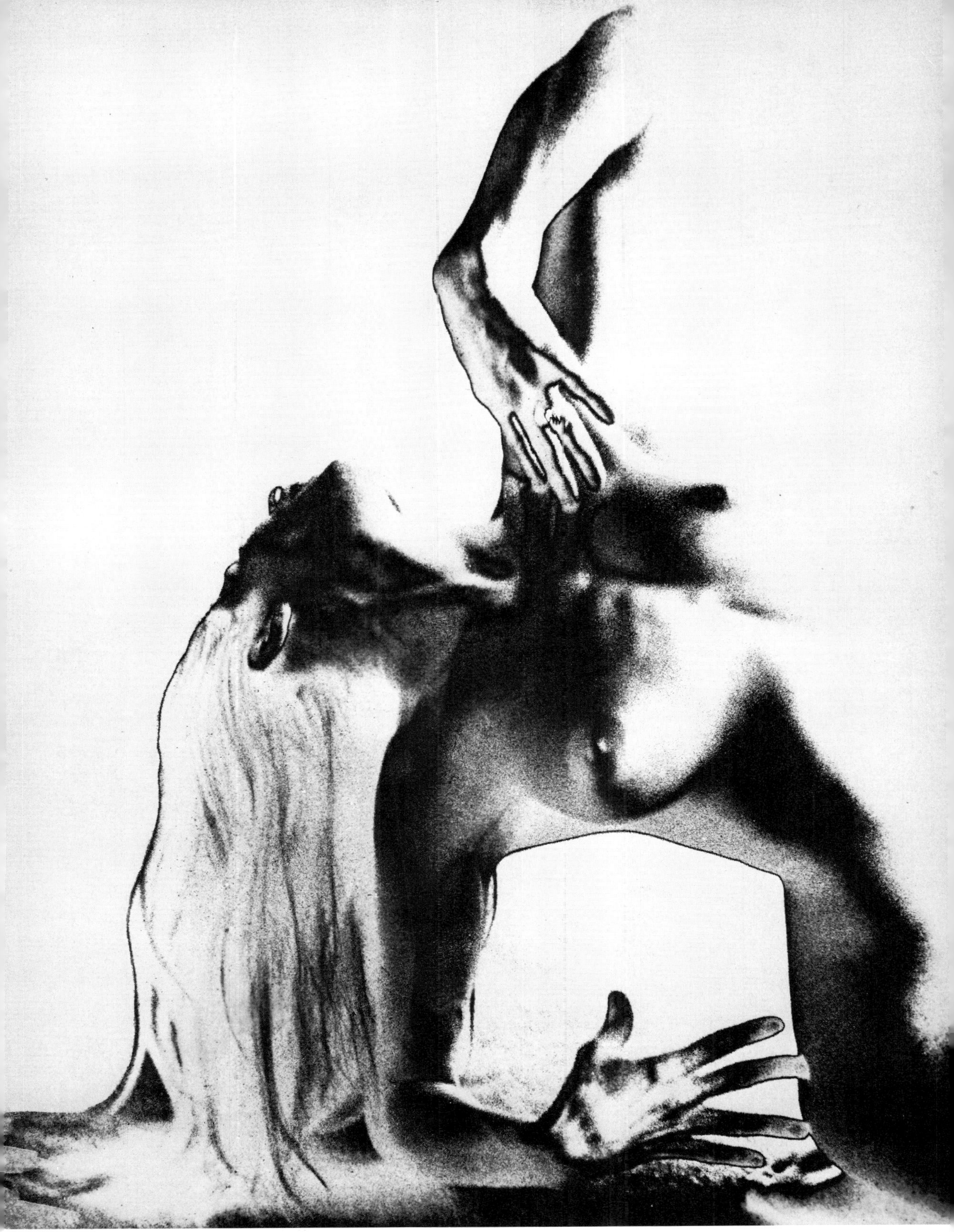

GP-2

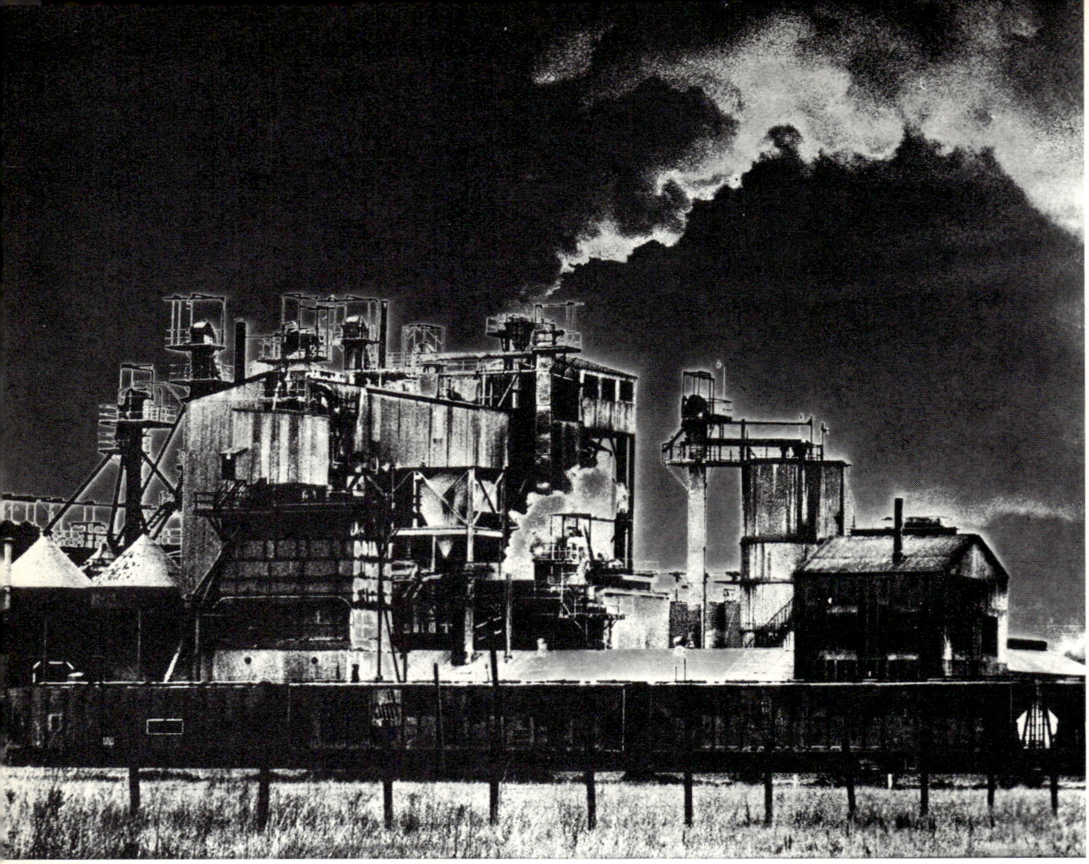

GP-3

GP-4

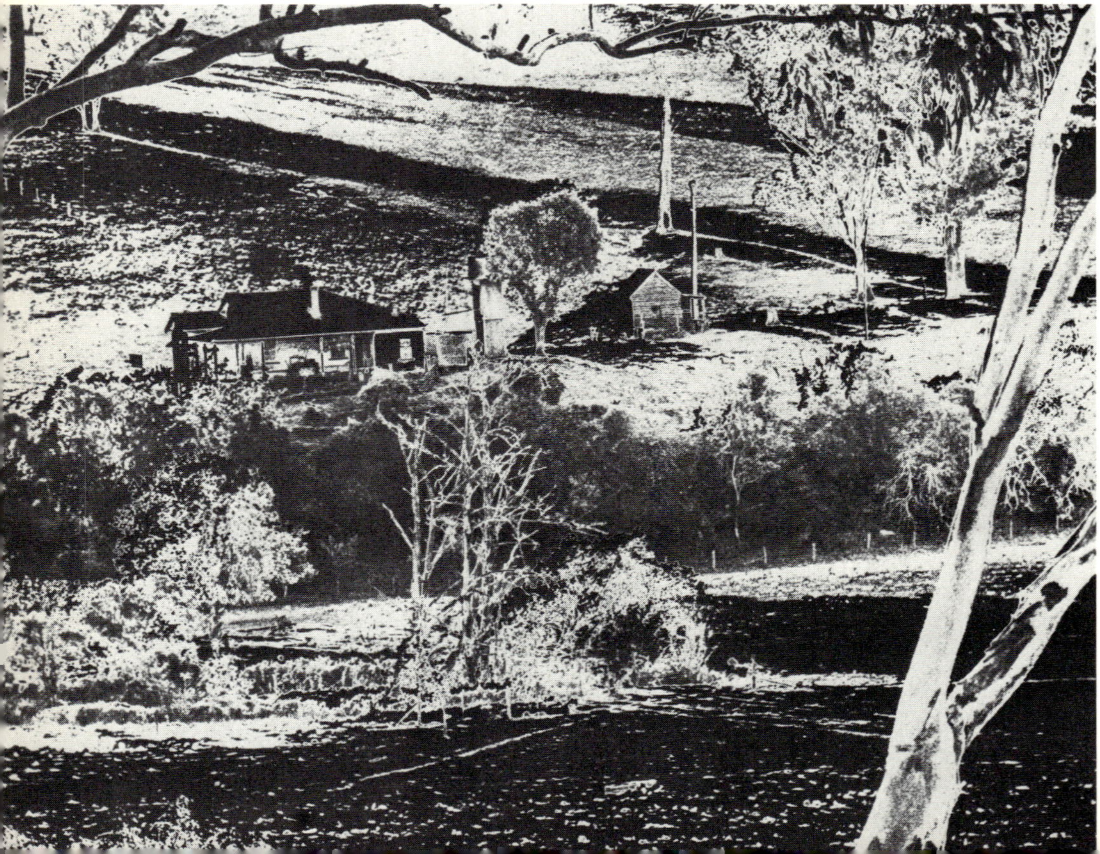

GP-5

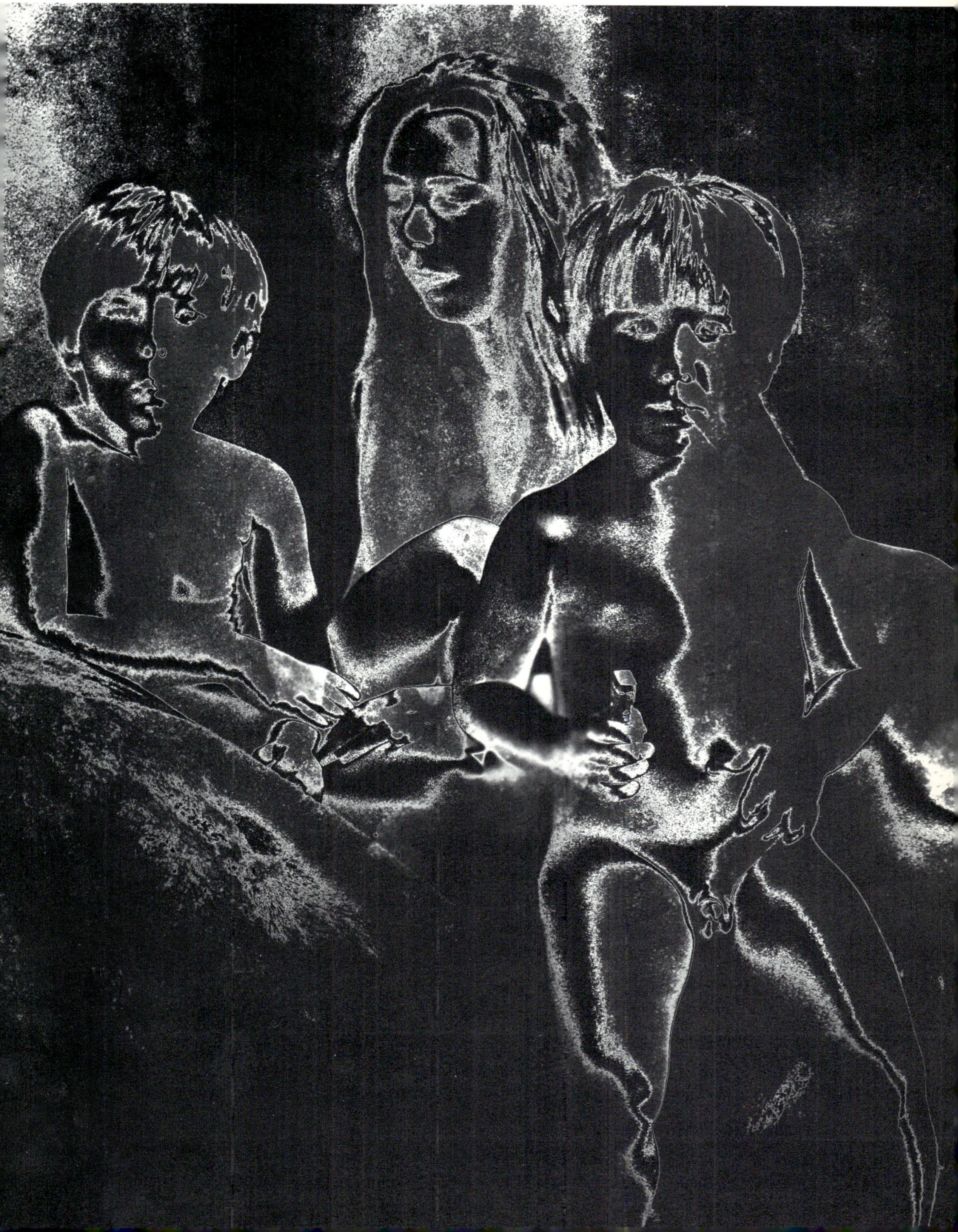

GP-6

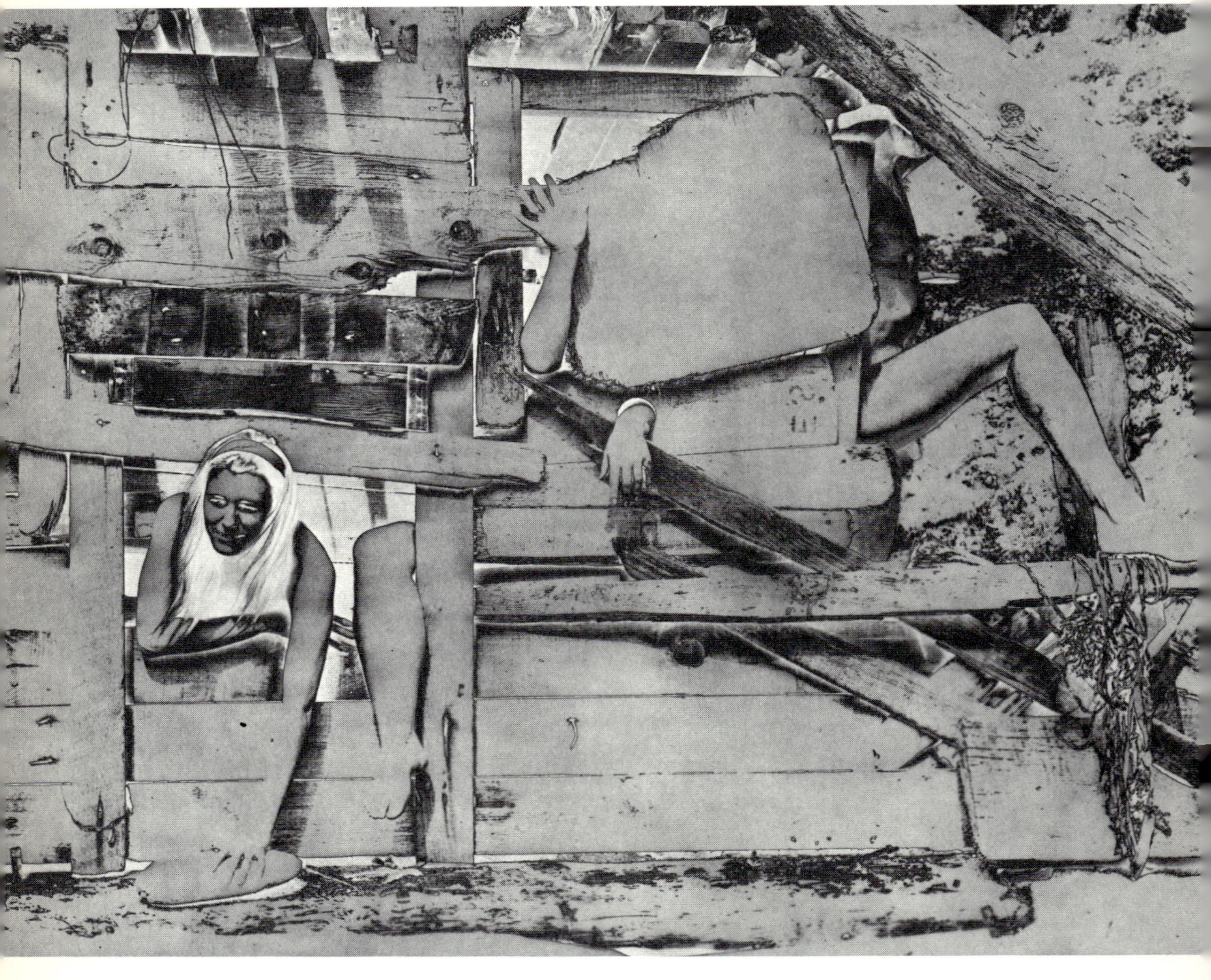

GP-7

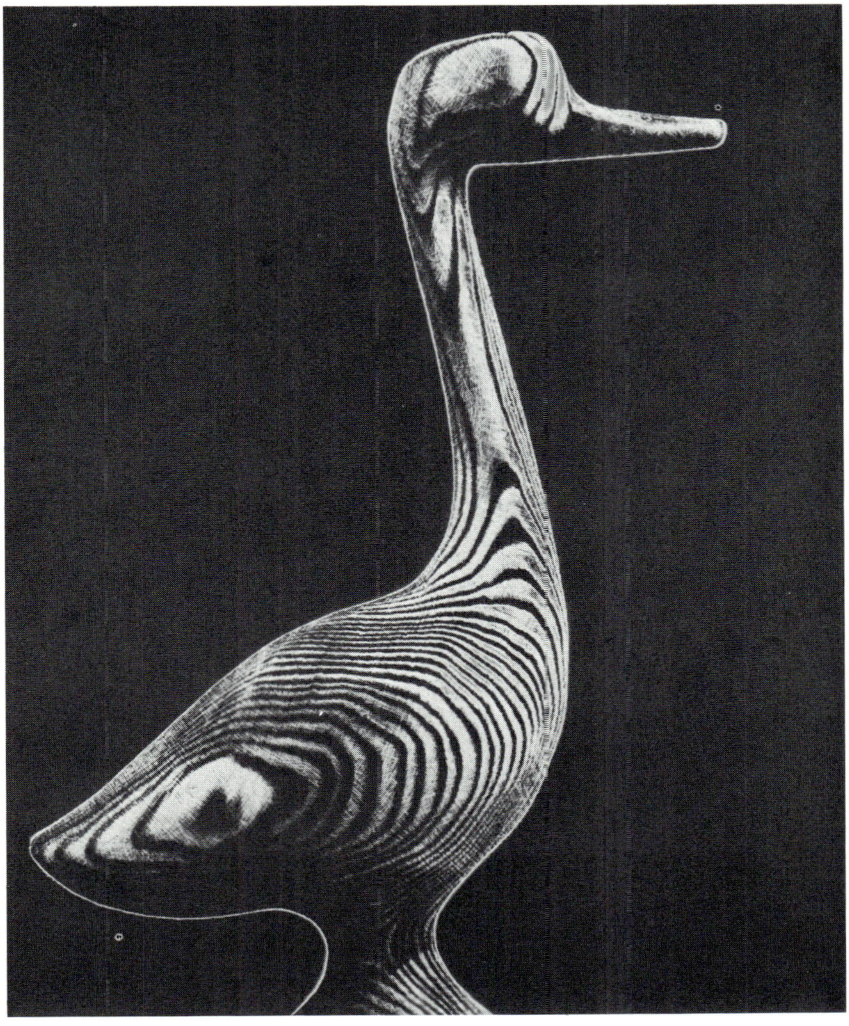

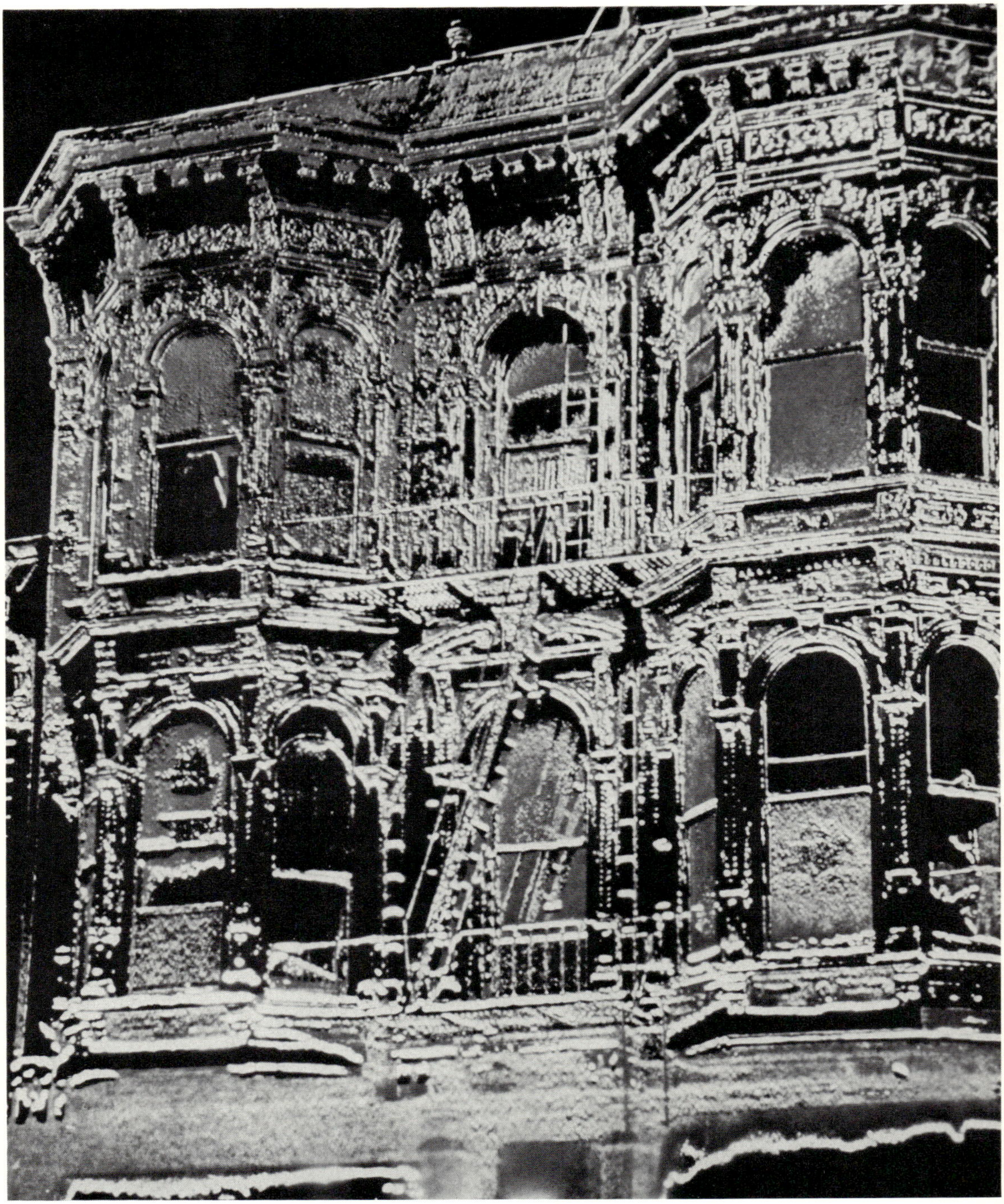

GP-8

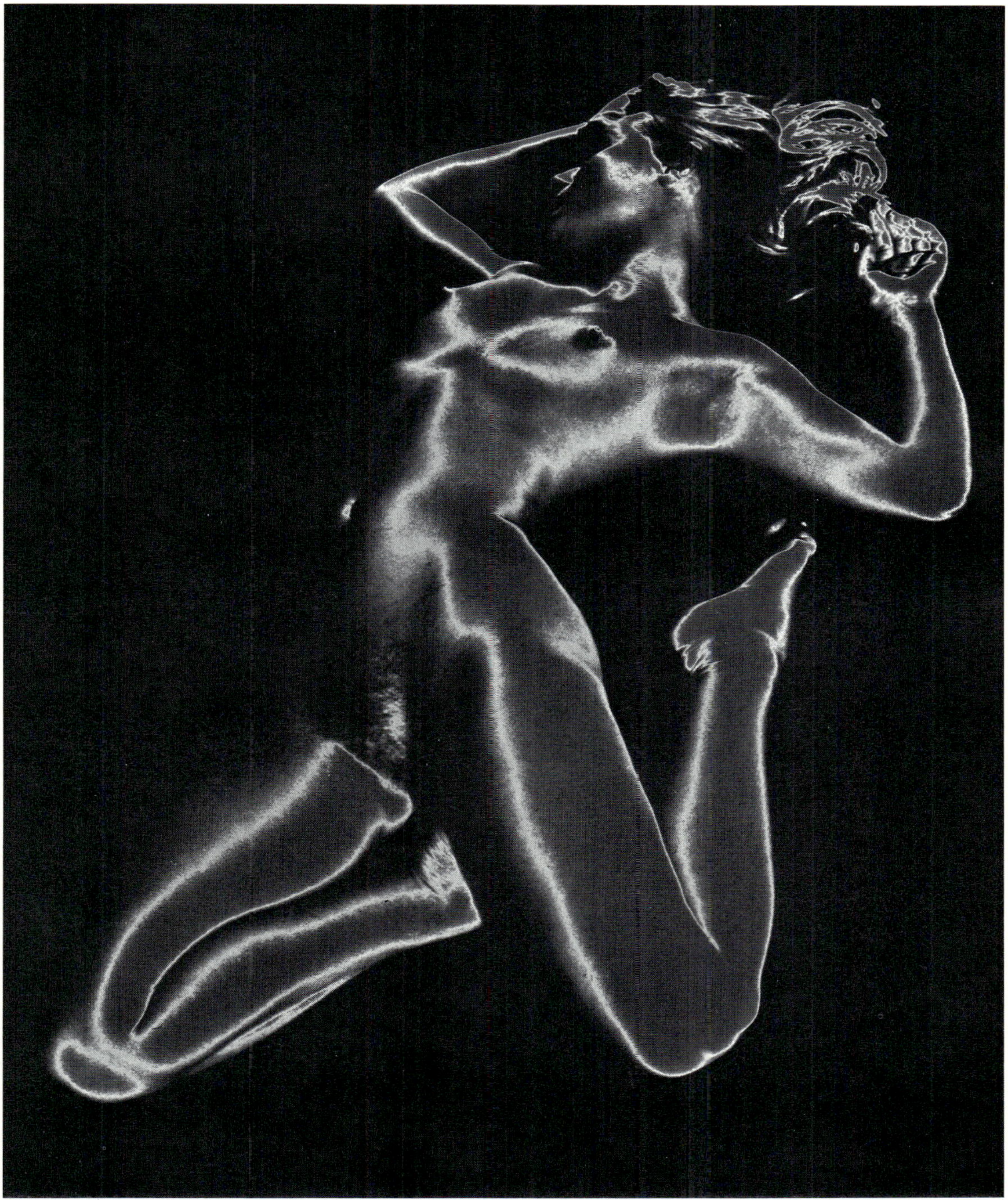

GP-9

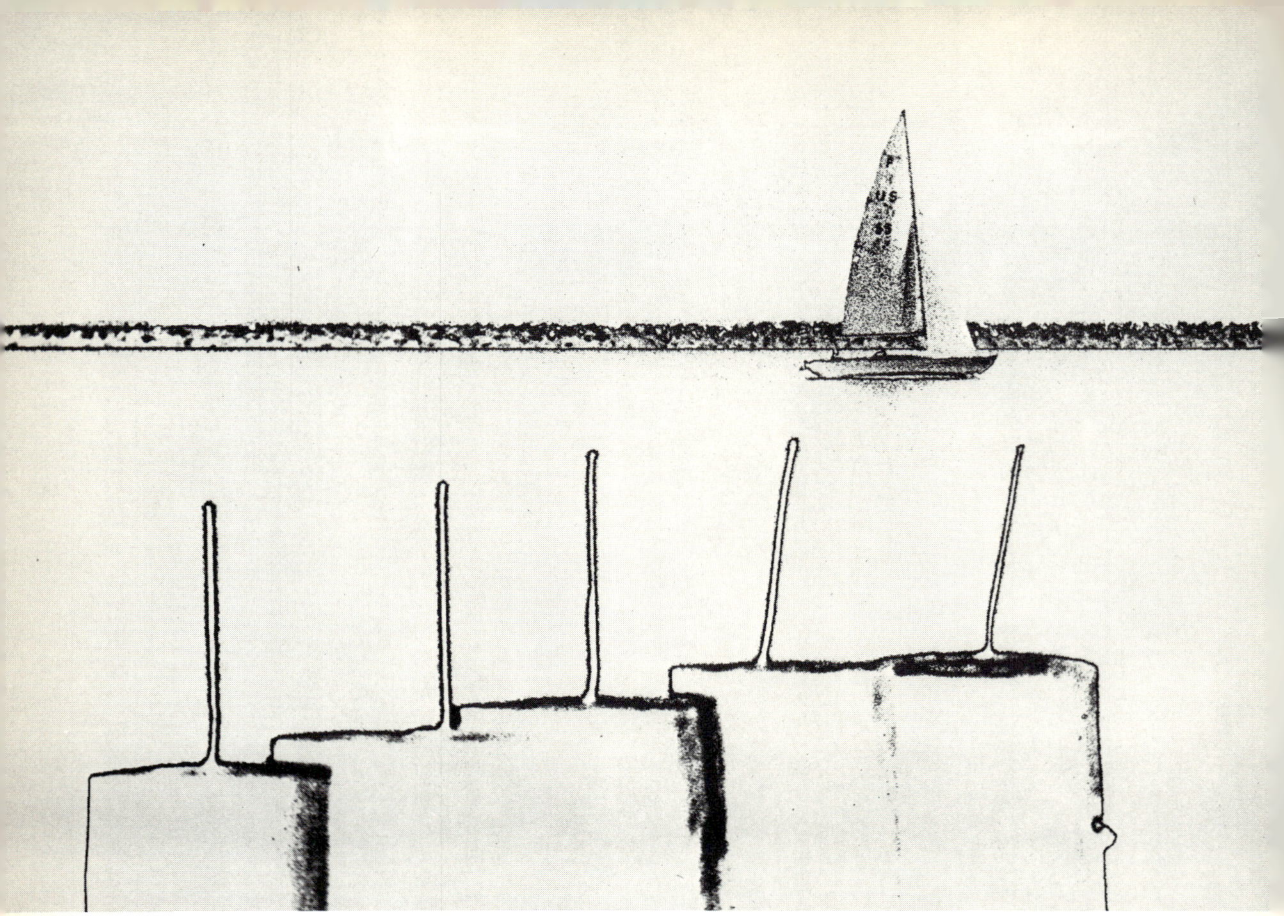

GP-10

GP-11

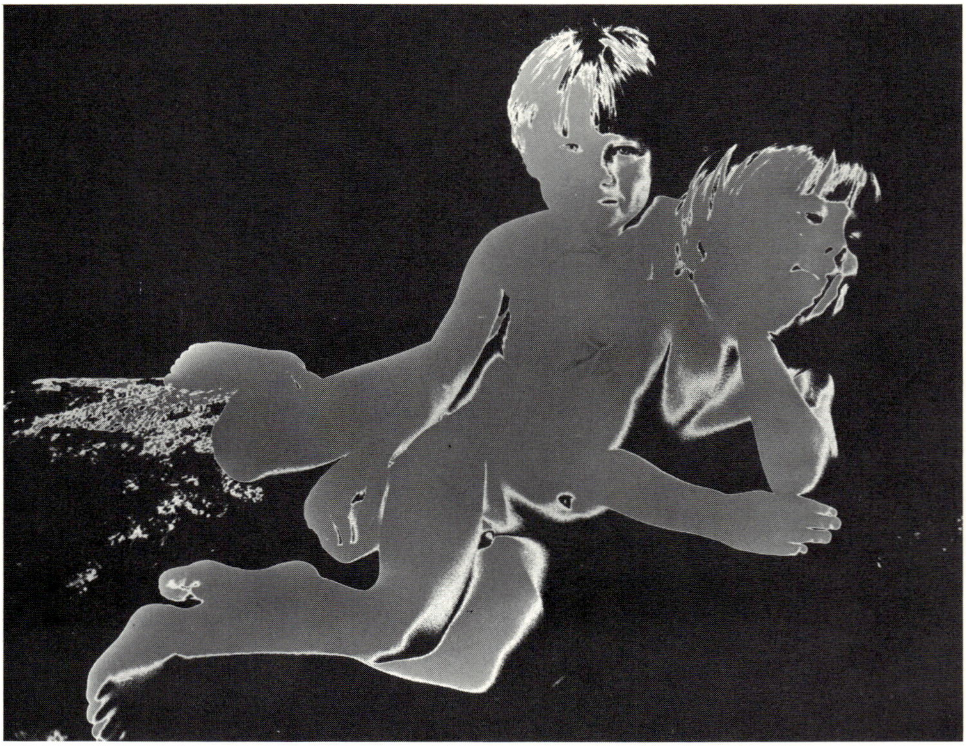

GP-12

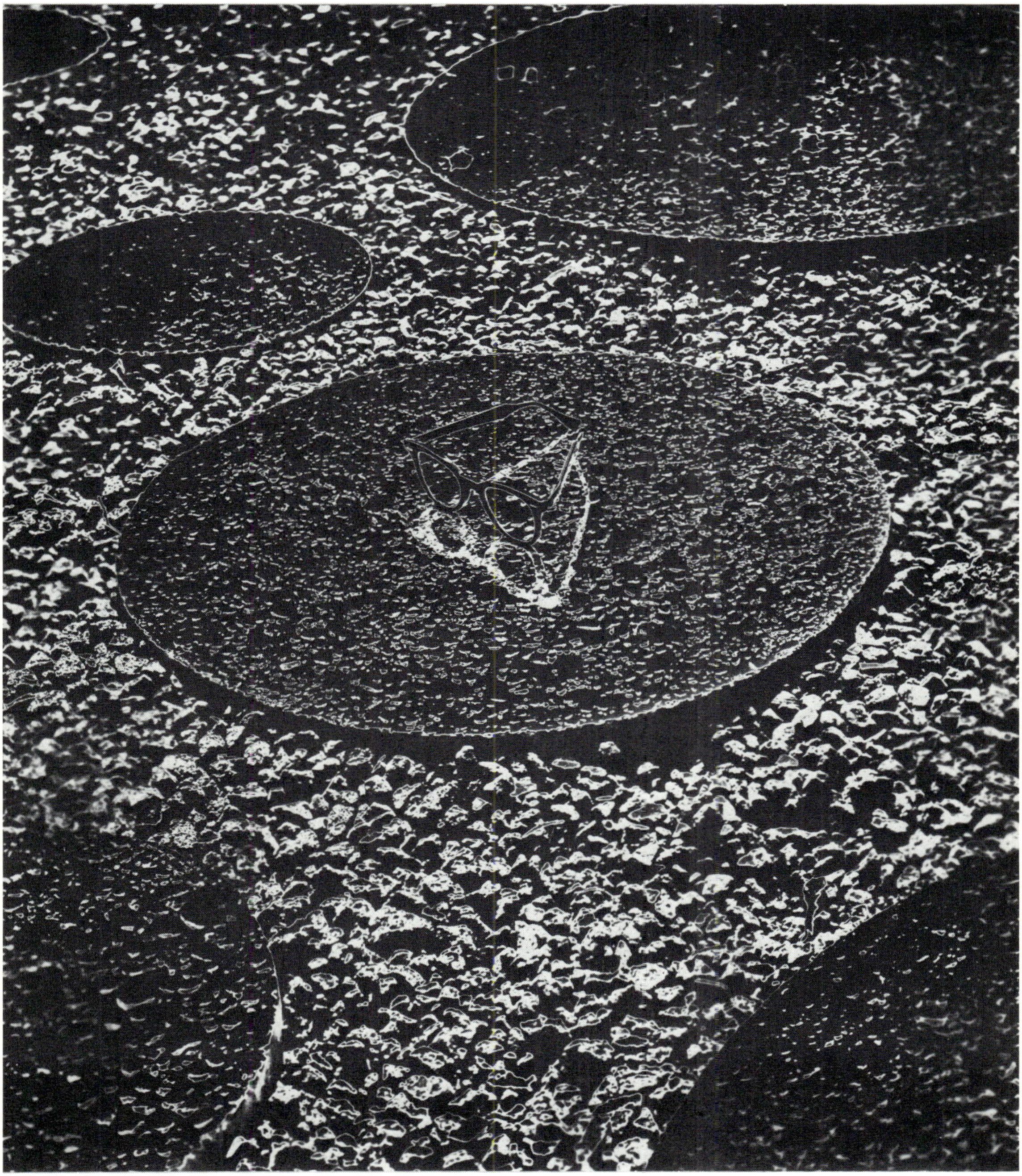

GP-13

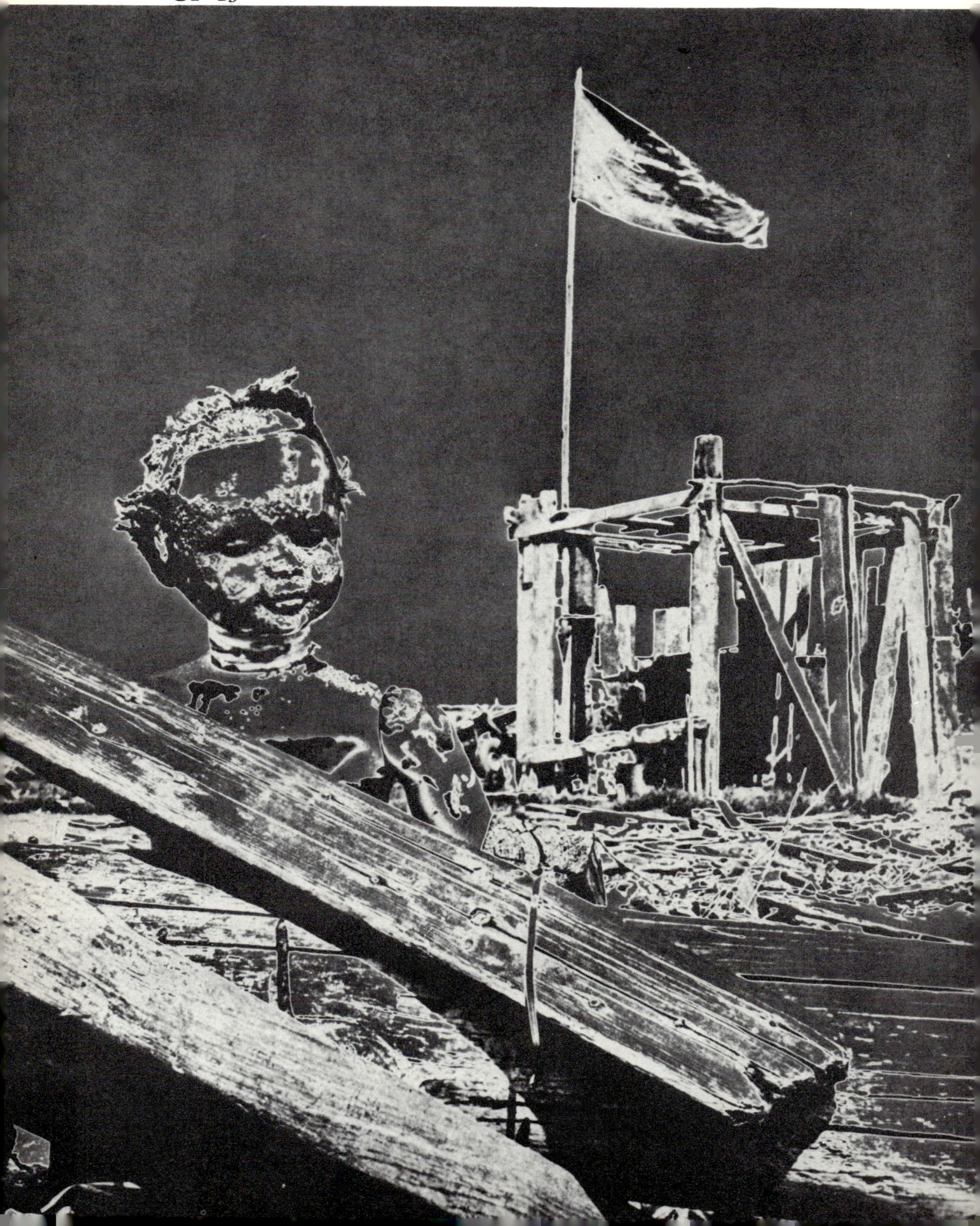

GP-14

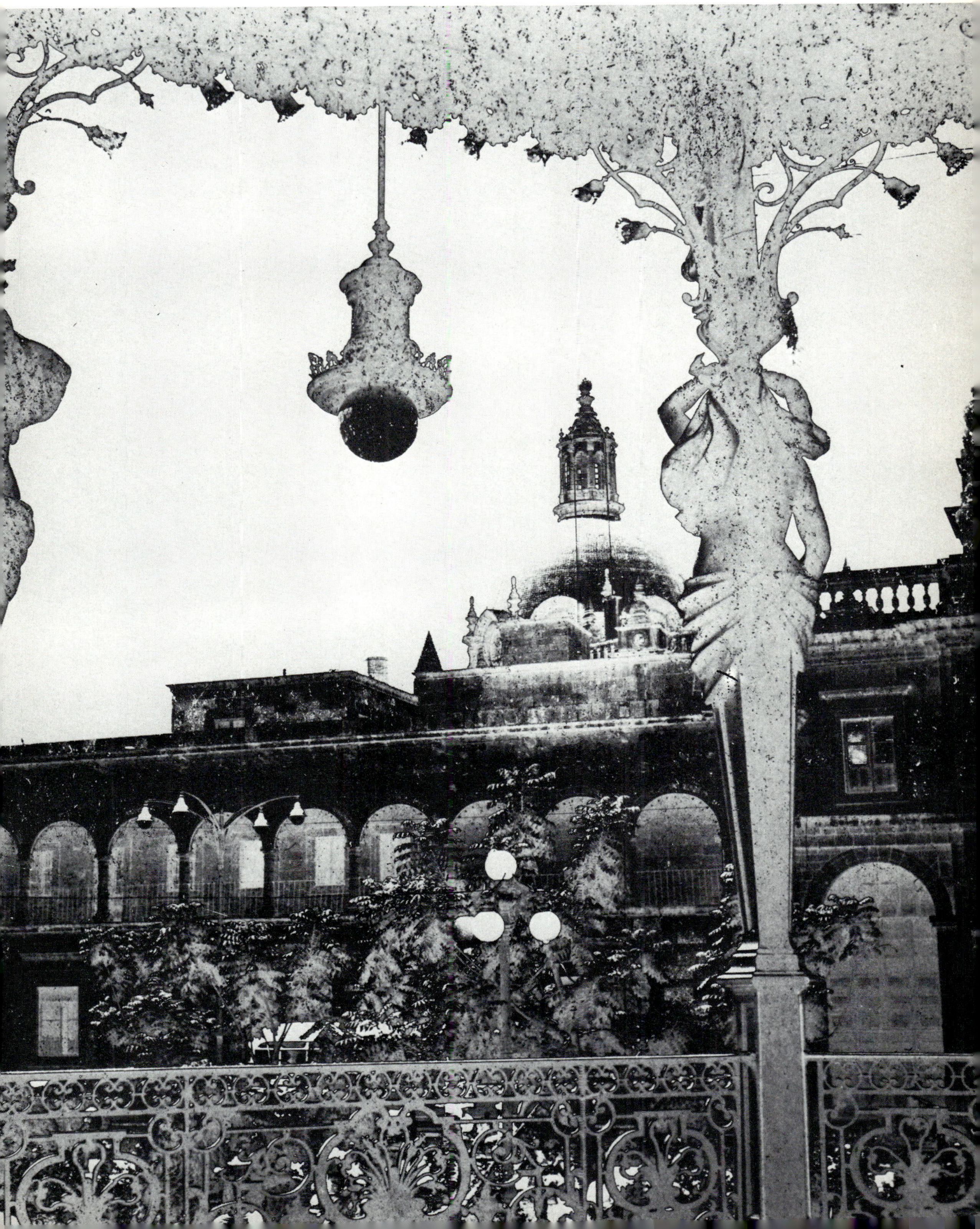

GP-16

GP-15

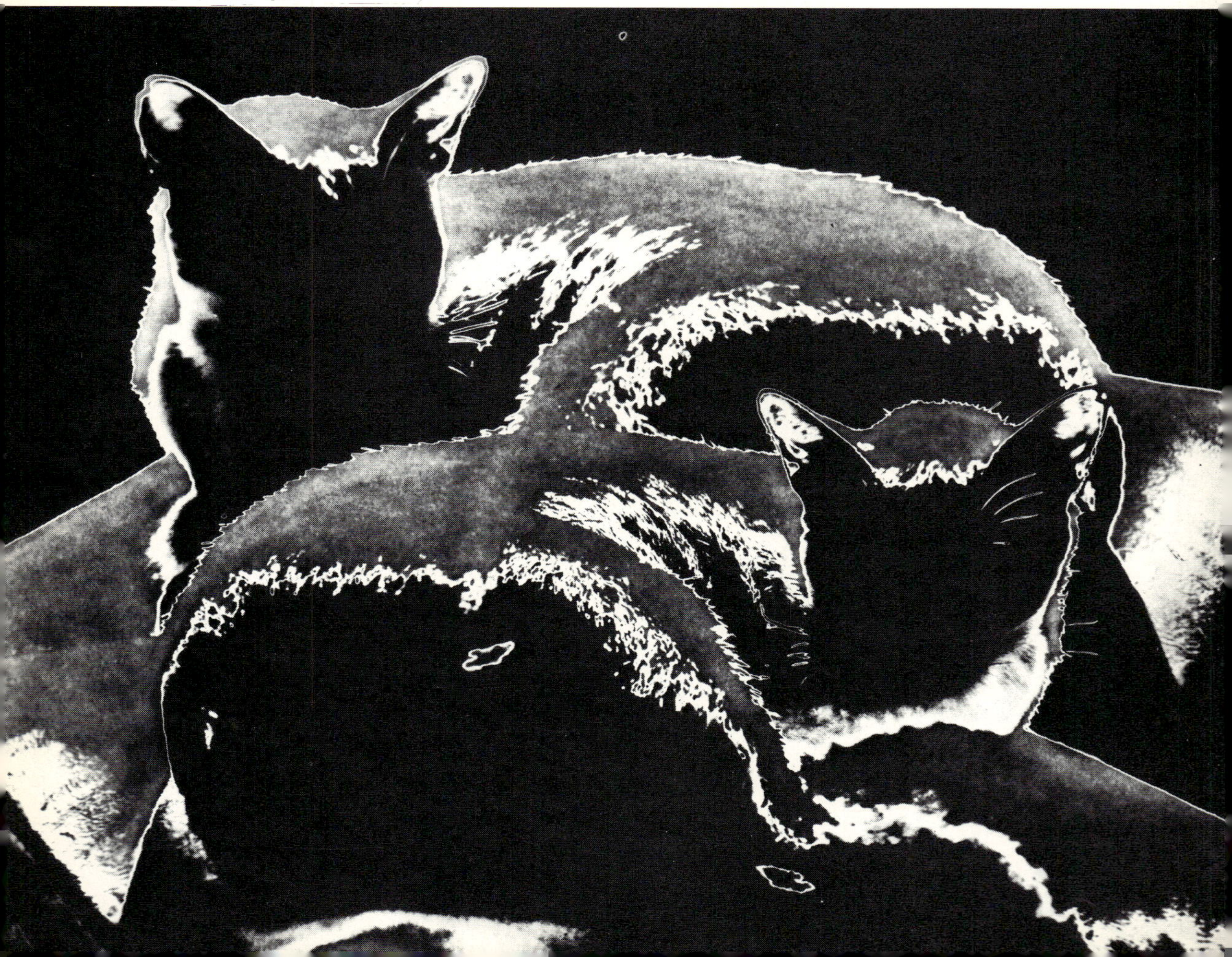

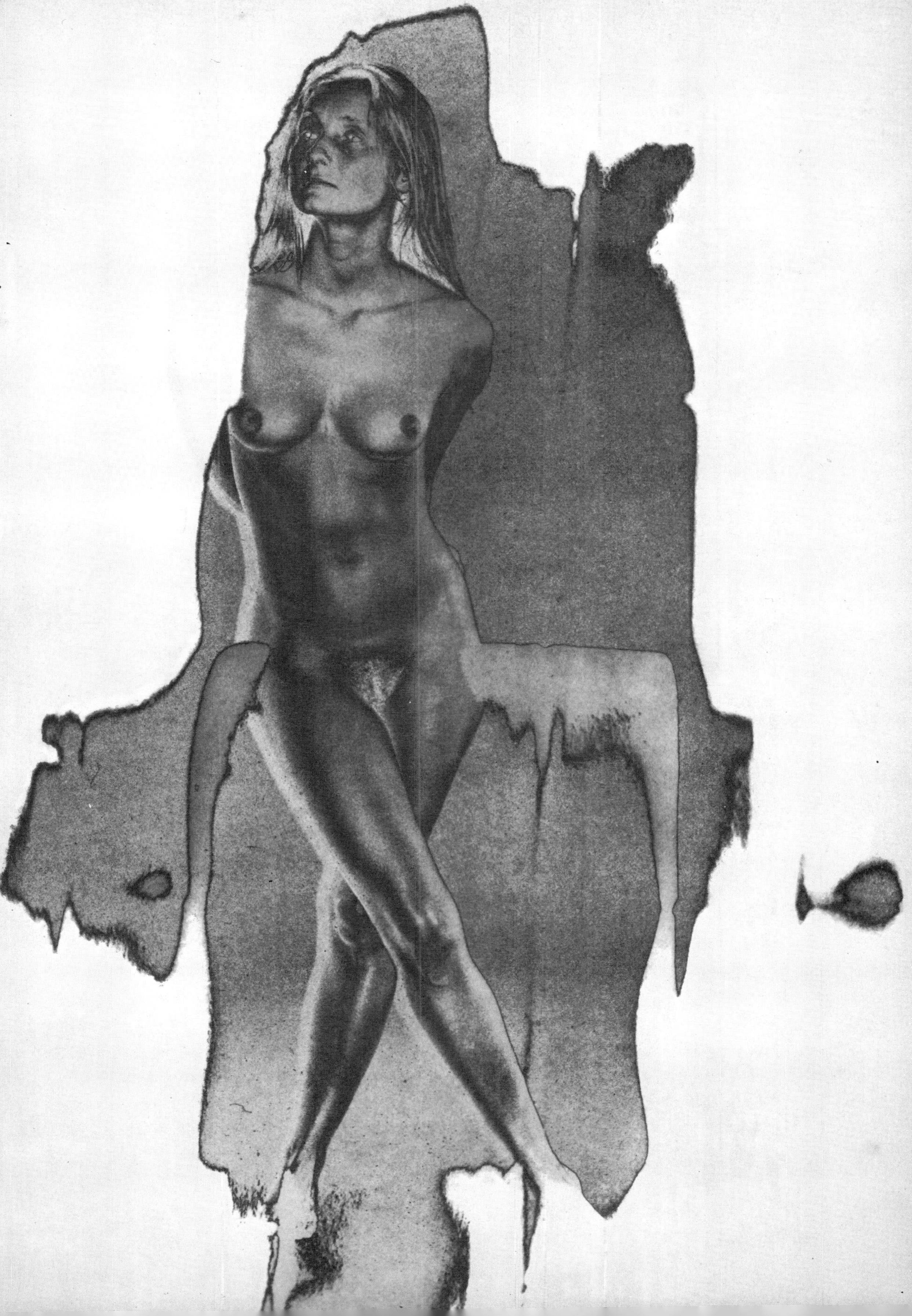

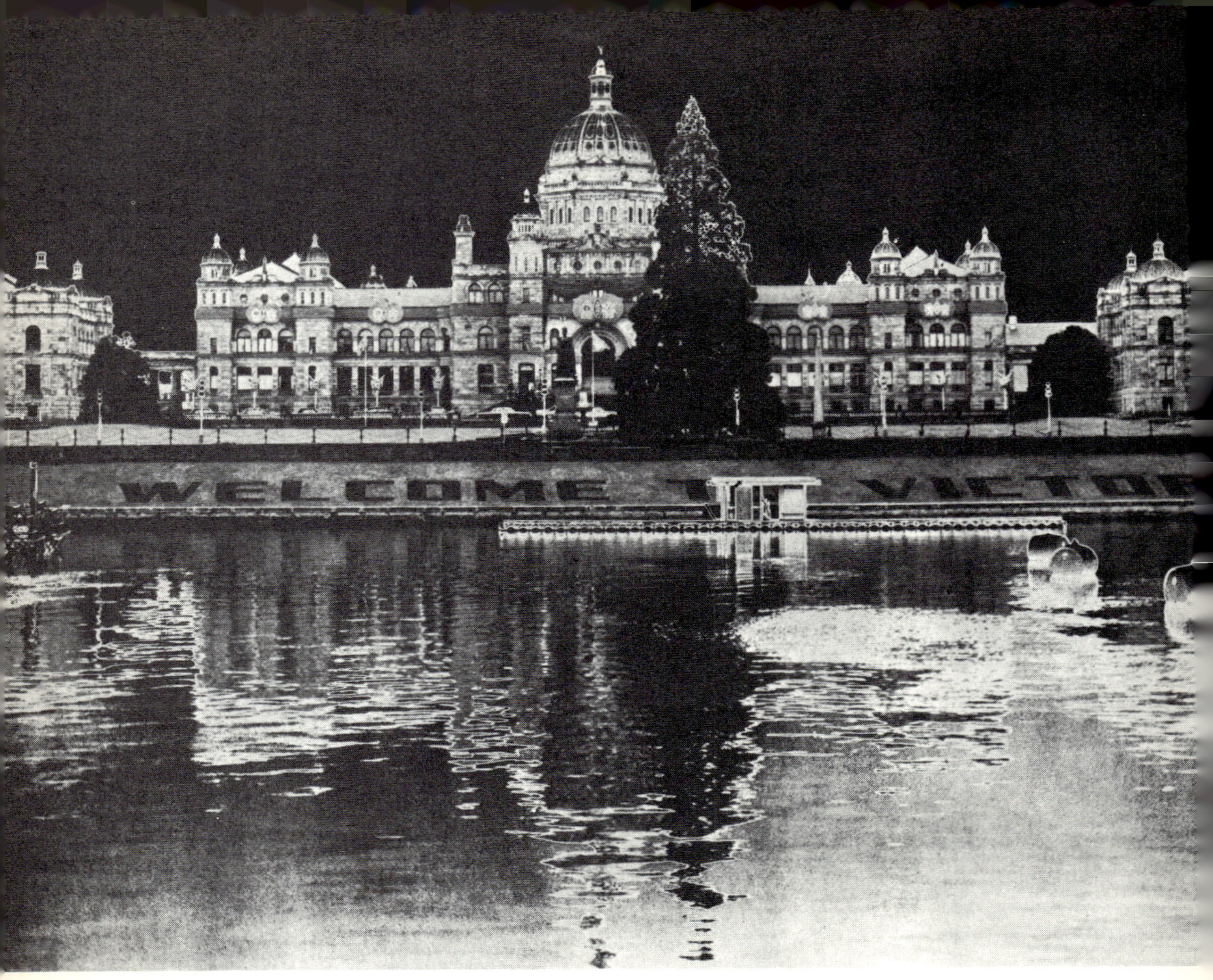

GP-17

GP-18

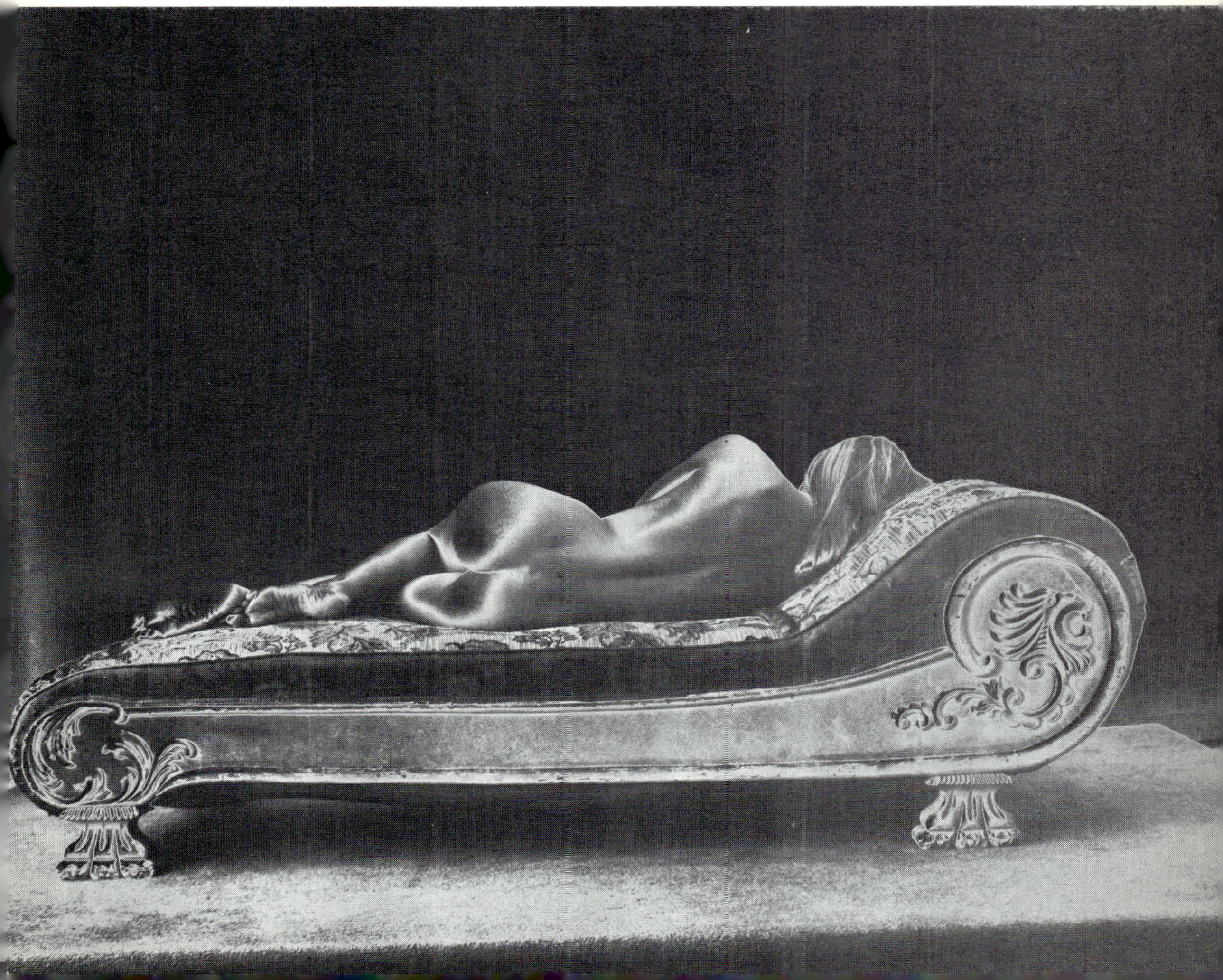

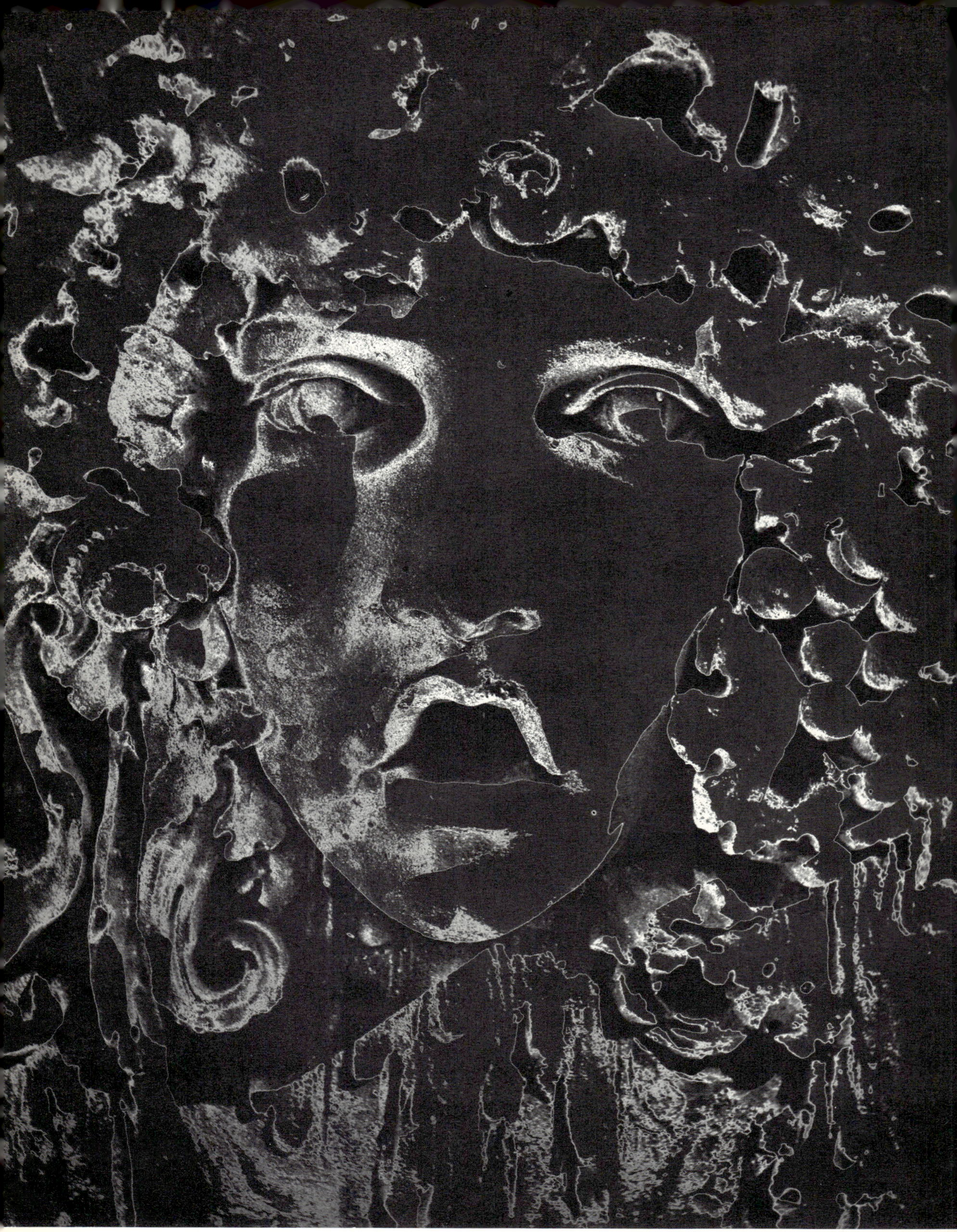

GP-19

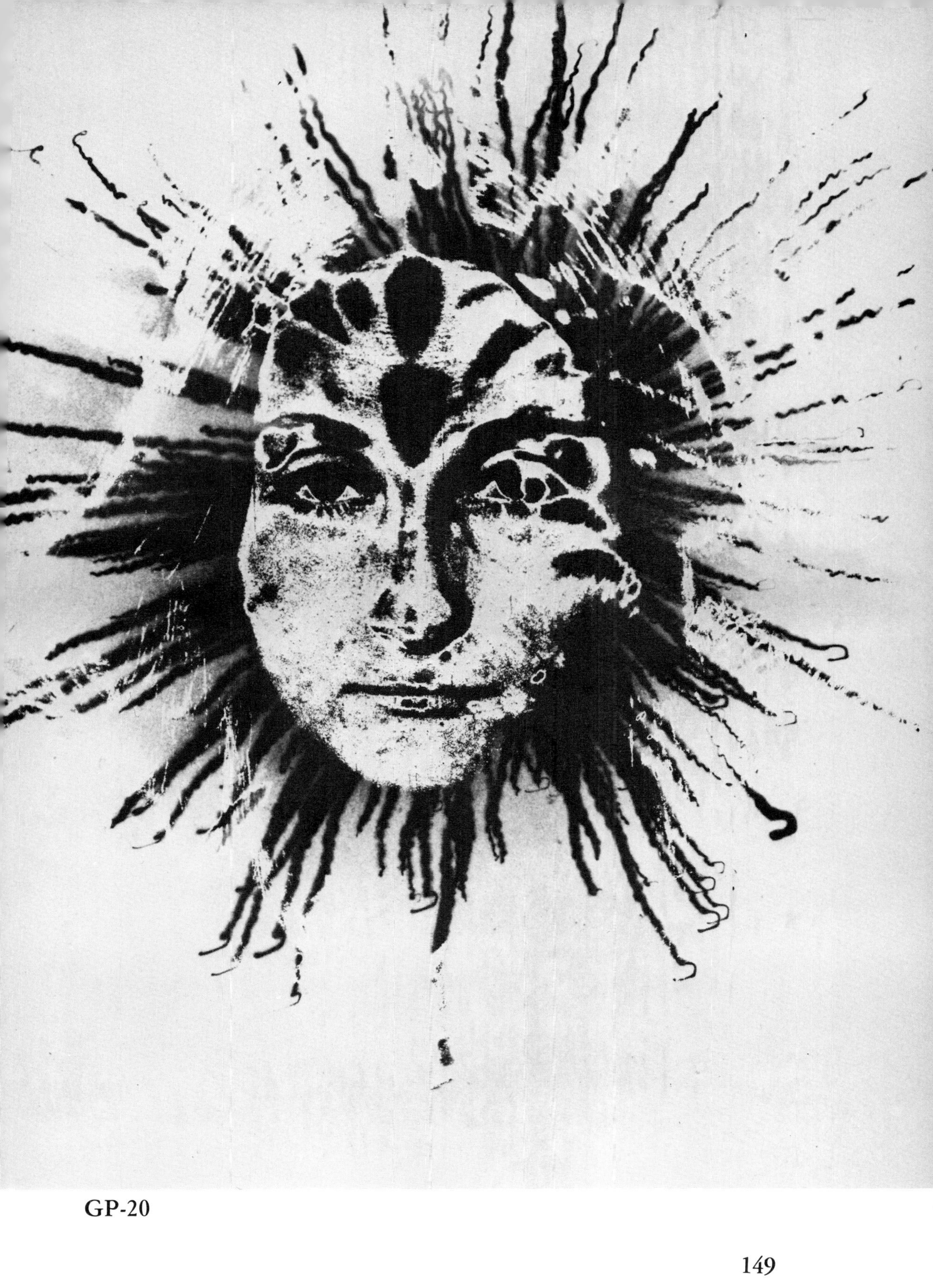

GP-20

GP-21

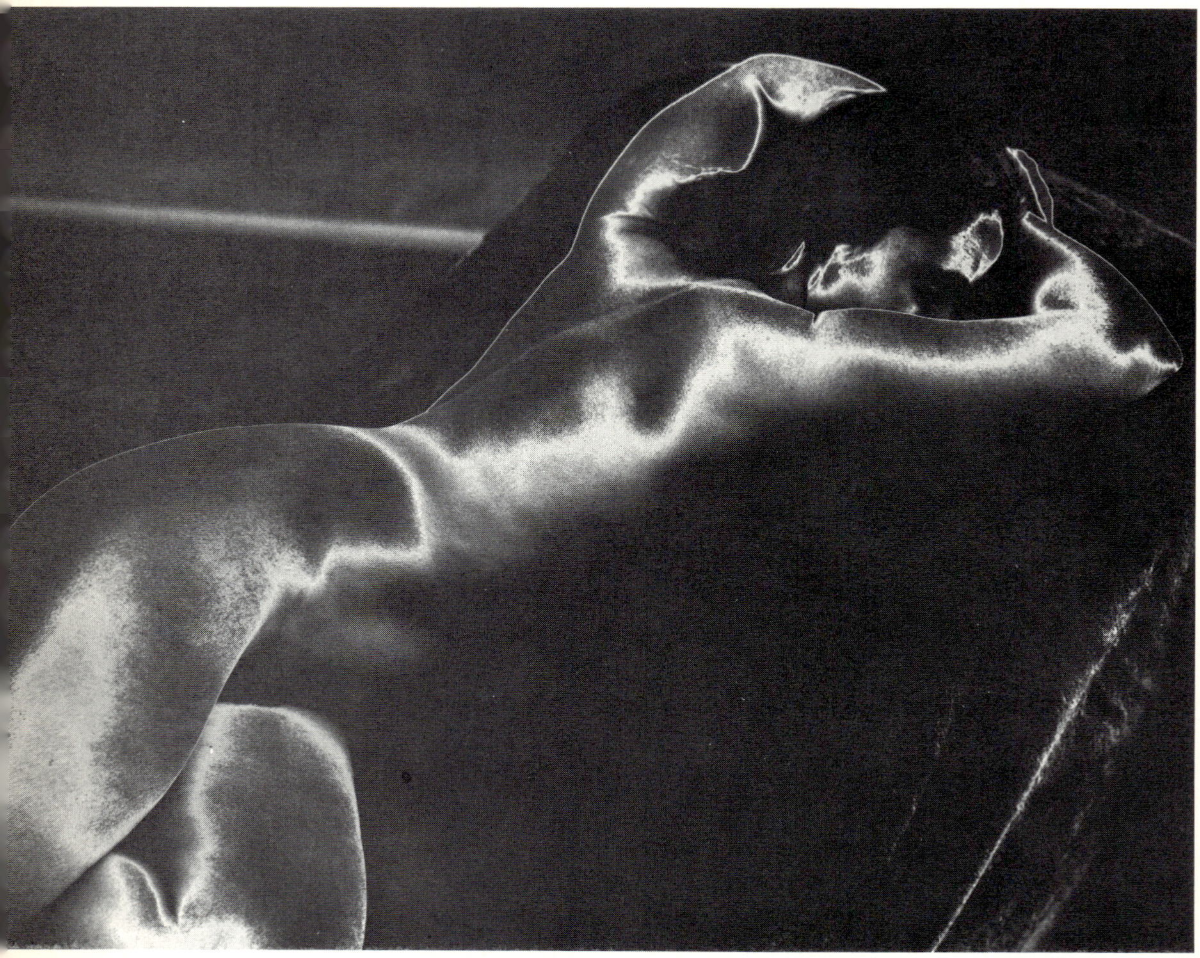

GP-22

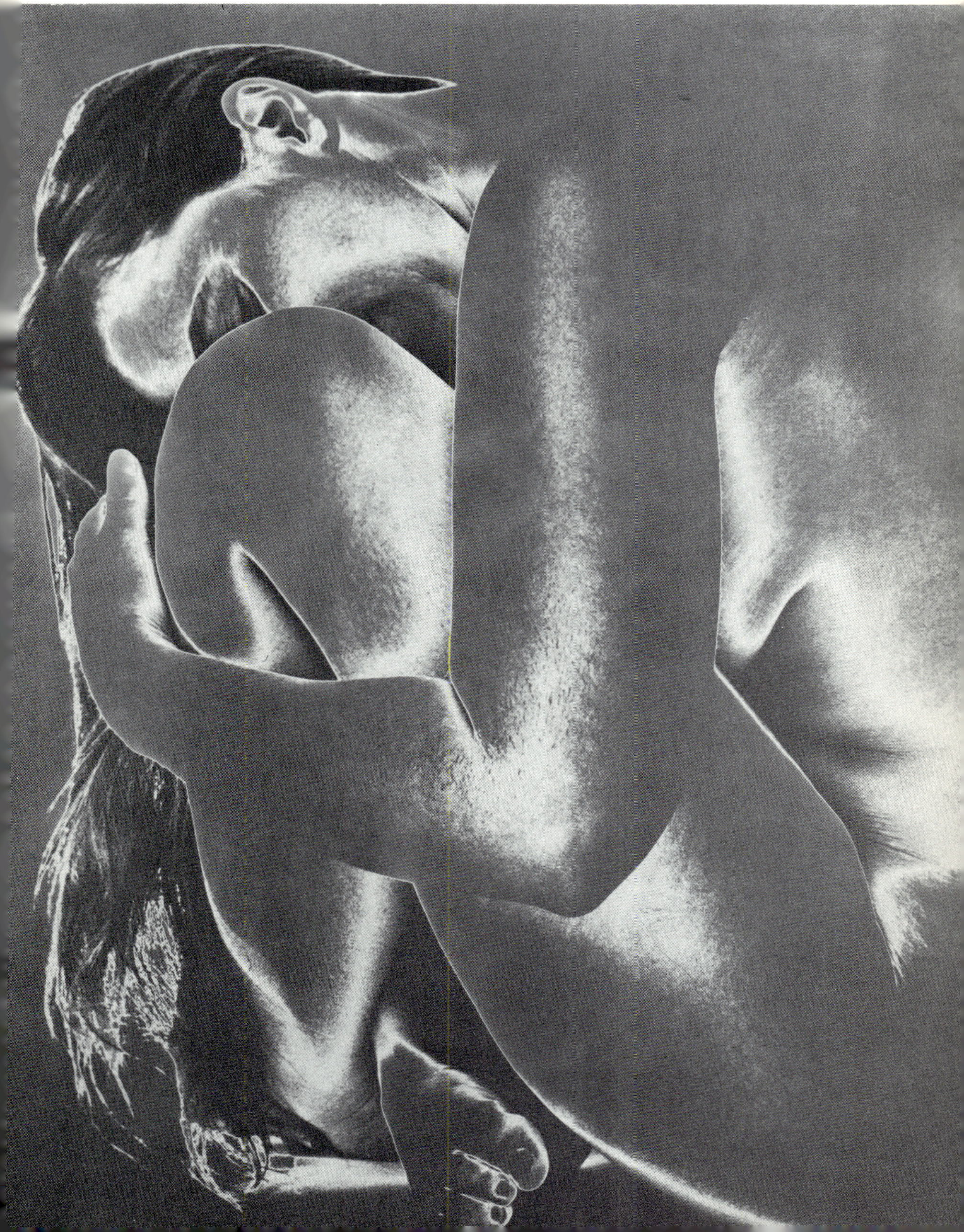

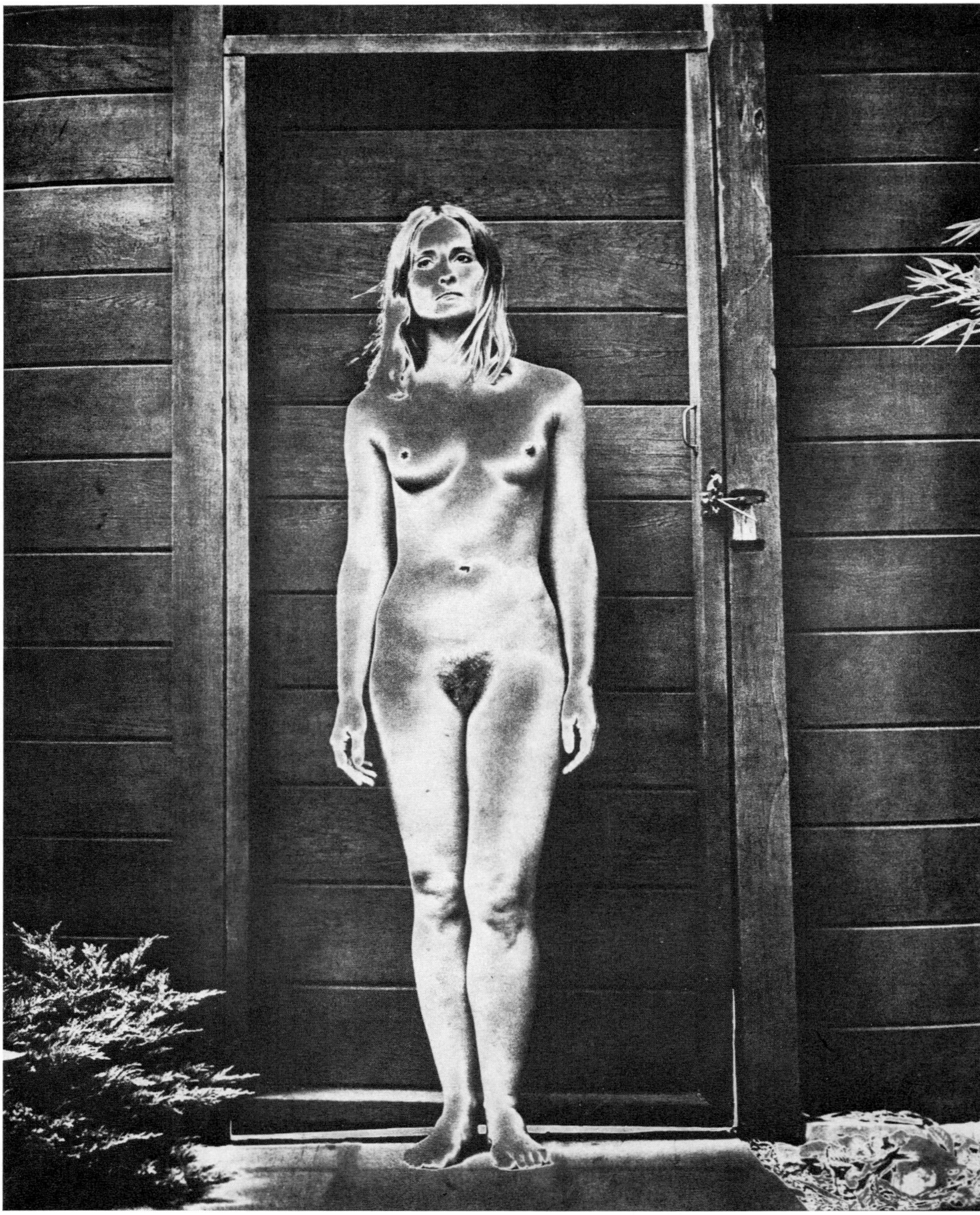

GP-23

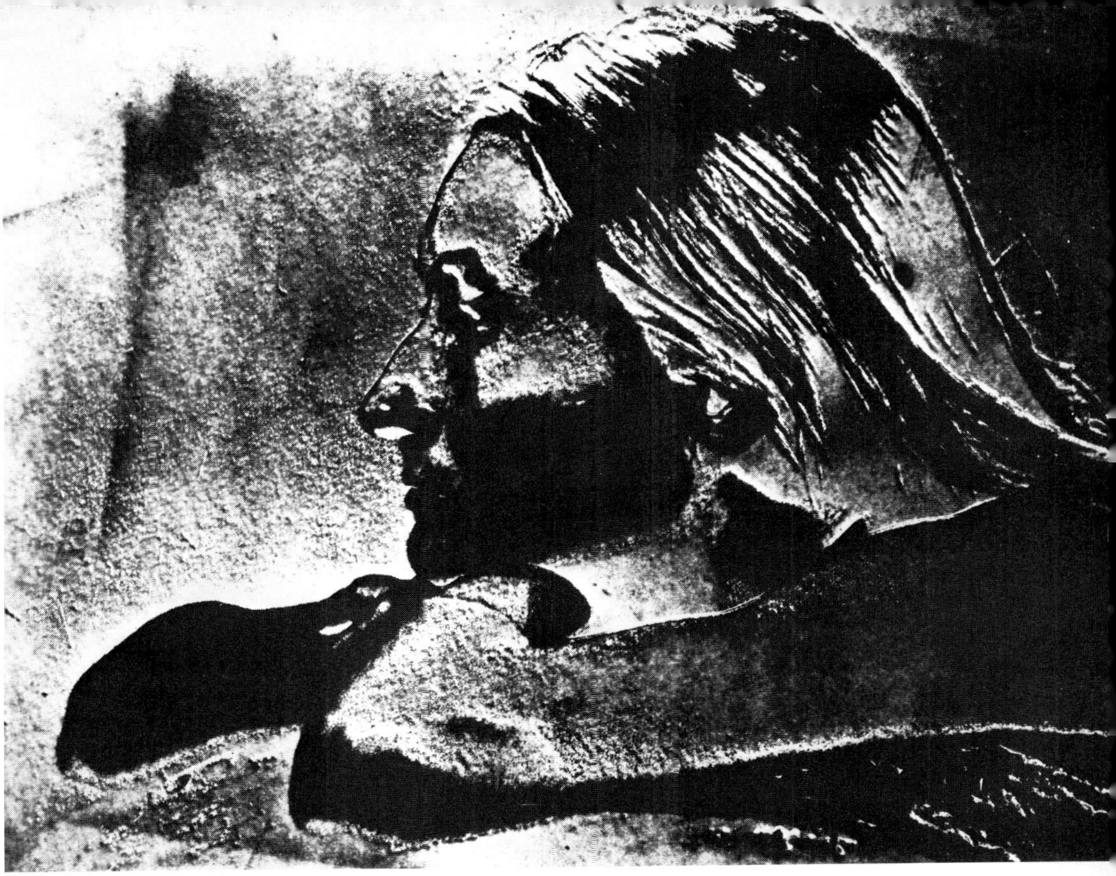
GP-24

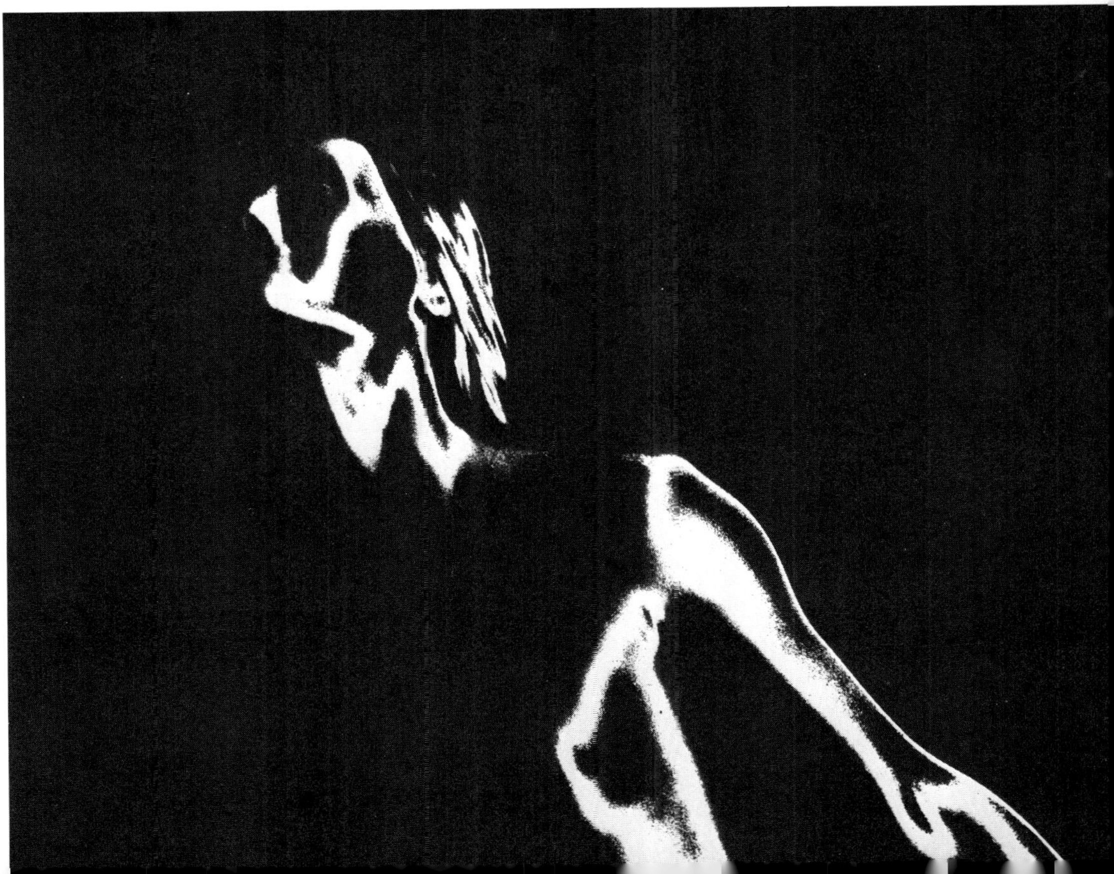
GP-25

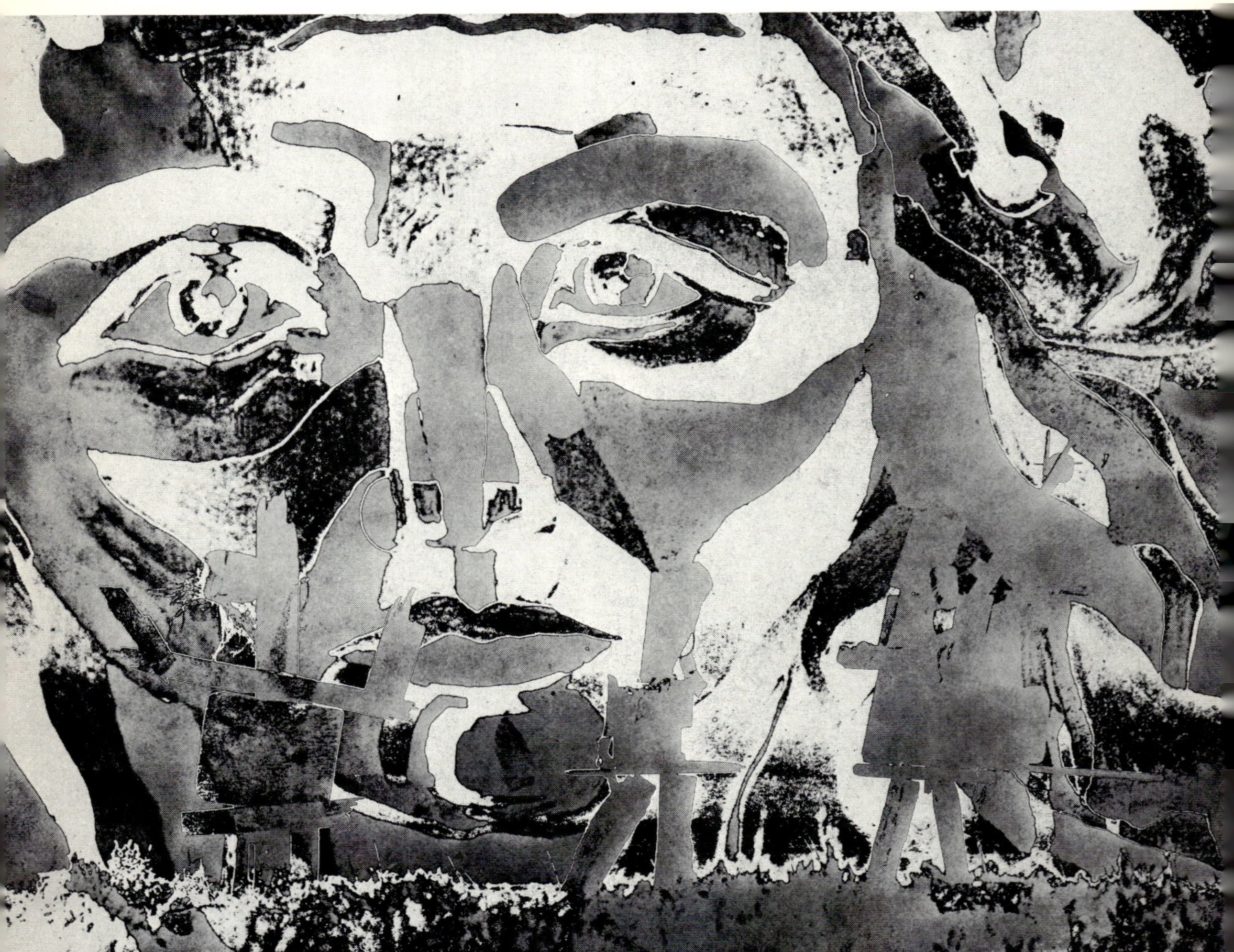

GP-26

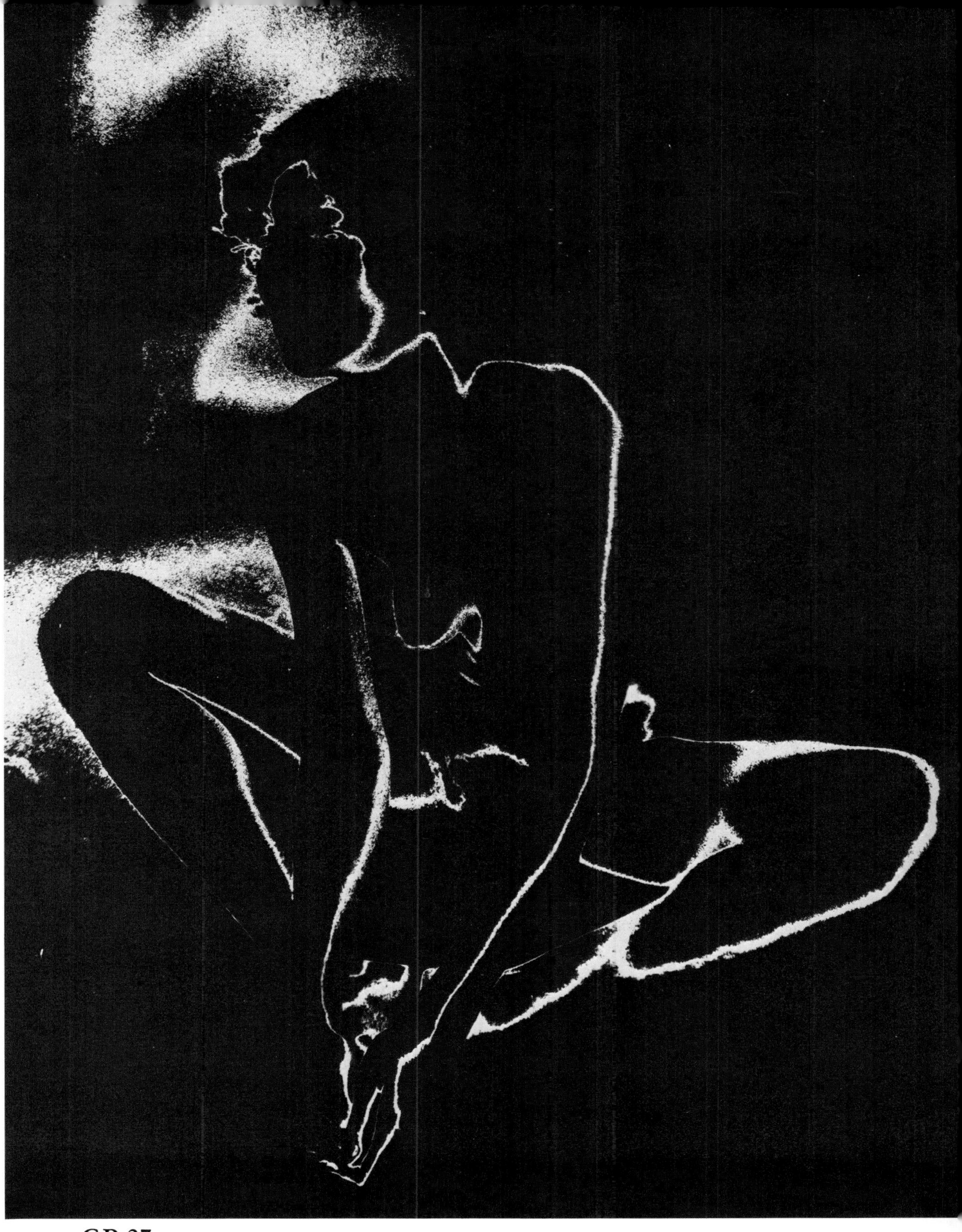

GP-27

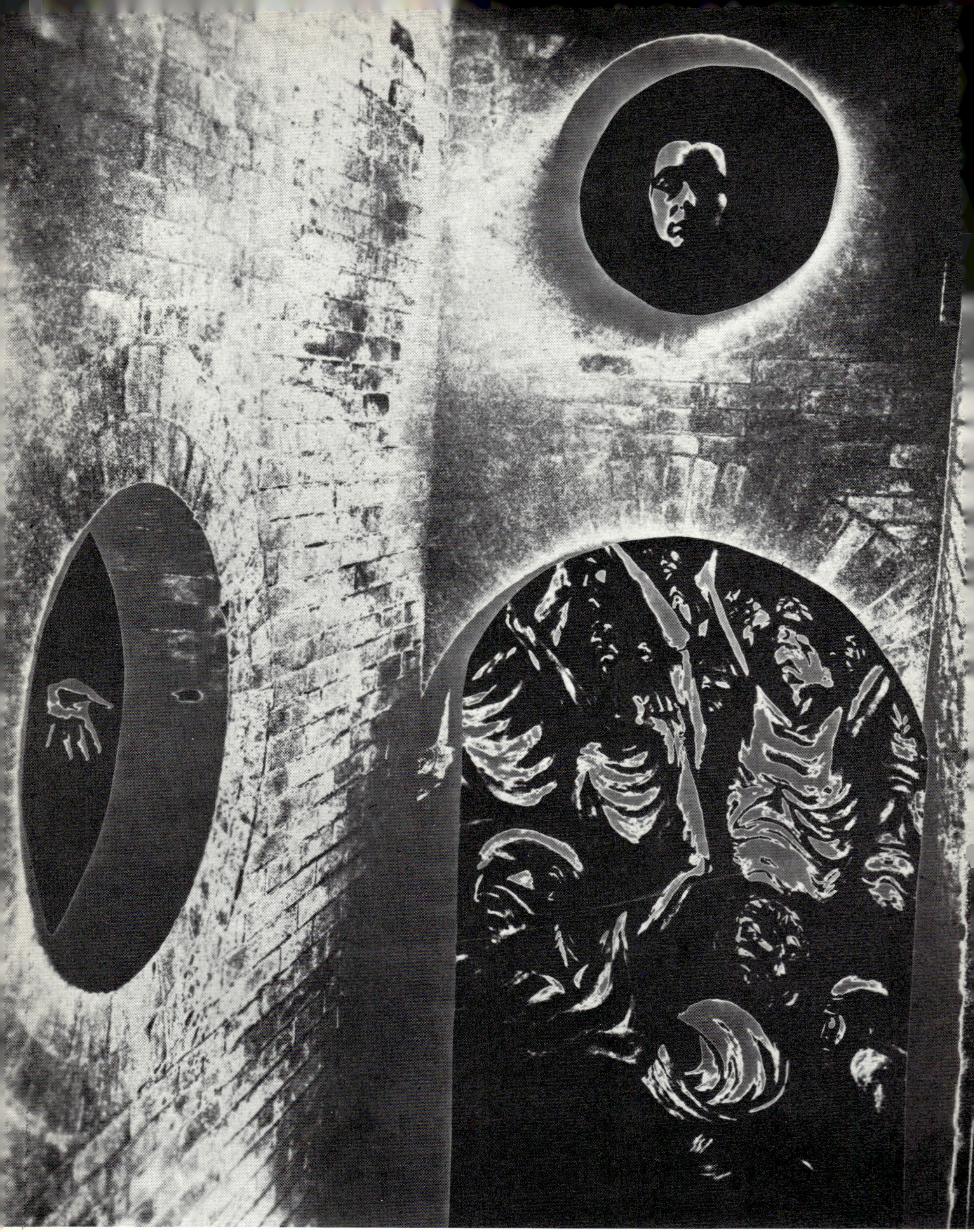

GP-28

Bibliography

Scientific and technical works relating to the Sabattier effect and solarization:

Barrows, R. S. and R. N. Wolfe, "Review Of Adjacency Effects In Silver Photographic Images." *Photographic Science and Engineering*, Vol. 15, No. 6 (November/December, 1971).

Berg, W. F., "New Cause Of Low-Intensity Reciprocity Failure." *Abridged Scientific Publications*, Eastman Kodak Research Laboratories, Vol. 29-30, Article No. 1163H (1947-1948).

———"The Photographic Subimage And Latent Image Dispersity." *Abridged Scientific Publications*, Eastman Kodak Research Laboratories, Vol. 33, Article No. 1255H (1951). This article discusses the sub-latent image (mentioned in our Chapter 5), and its relation to exposure and development.

Brothers, A., *A Manual Of Photography*, Philadelphia, J. B. Lippincott Co., 1892. This author attributes pseudo-solarization entirely to optical shielding.

Chibisov, K. V., M. V. Fok, E. A. Galashin, and E. P. Senchenkov, "The Nature Of Photographic Sensitivity And The Mechanism Of Formation Of The Latent Image." *The Journal of Photographic Science*, Vol. 21, No. 3 (May/June

1973). This paper, which was published after this book was written, supports a theory which views the photographic emulsion grain as a super-saturated solid solution of silver atoms in the silver halide crystal. The action of light is to initiate the coagulation of these atoms into silver particles of colloidal size, consisting of tens or hundreds of atoms. It is these colloidal particles which constitute the latent image in the emulsion. This differs from the traditional view in which each latent-image speck contains only a few atoms of silver. According to the new theory, the smaller of these colloidal silver particles are dissolved during the first few seconds of development, while the larger ones grow to developable size. If this view of the photographic process is substantiated by future experiments, it could play an important part in the explanation of many reversal phenomena, including the Sabattier effect.

Encyclopedia of Photography, The, Vol. 18, Greystone Press, 1967, p. 3429.

Encyclopedic Dictionary of Physics, Vol. 5, Pergamon Press, 1962, p. 405. This consists chiefly of an elaborate table of known exposure effects and their variations.

Friedman, J. S., *American Photography*, July, 1944, particularly p. 41. Discusses latent-image theory and true solarization.

Langford, M., *Advanced Photography*, New York, Focal Press, 1969.

Marriage, A., "A Note On The Sabattier Effect." *The Photographic Journal*, (November, 1937), p. 619. A study that attributes the Sabattier effect mainly to optical shielding.

Mees, C. E. K. and T. H. James, *The Theory of the Photographic Process*, 3rd ed. New York, The MacMillan Company, 1966. The best general reference on the scientific aspects of the subject (particularly Chapter 8). Includes an extensive bibliography.

Mott, N. F. and R. W. Gurney, *Electronic Processes In Ionic Crystals*, New York, Dover, 1964. Chapter VII is entitled "Photochemical Processes in Silver Halides and the Photographic Latent Image." This is the classic work on the theory of the latent image. Still one of the best descriptions in print.

Neblette, C. B., *Photography, Its Materials And Processes*, New York, Van Nostrand Reinhold Company, 1962. Chapter 12 gives a good review of the theory of latent-image formation. Chapter 15 briefly reviews the important reversal effects and their possible causes. Chapter 16 discusses theory of development and distinguishes chemical and physical types.

Stevens, G. W. W. and R. G. W. Norrish, "Border Effects Associated With Photographic Reversal Processes." *The Photographic Journal*, (January, 1937), p. 20, and "The Mechanism Of Photographic Reversal." *The Photographic Journal*, (August, 1938), p. 513. These two articles report on what is probably the most thorough investigation into the nature and causes of the Sabattier and border effects.

Todd, H. D. and R. D. Zakia, *Sensitometry*, Hastings-on-Hudson, Morgan   Morgan, 1969.

Trivelli, A. P. H., "The Solarized Latent Image of Photographic Emulsions." *Abridged Scientific Publications*, Eastman Kodak Research Laboratories, Vol. 29-30, Article No. 416 (1947-1948).

Popular accounts of pictorial uses of the Sabattier effect:

"Darkroom Pictures." *Popular Photography*, (May, 1948), p. 60.

Fehr, G., "What Is Solarization?" *Popular Photography*, (October, 1949), p. 68.

Geraci, J., "How To Solarize Prints." *American Photography*, (September, 1952), p. 20.

Godsey, T., "Simple Experiments In Solarization." *Popular Photography*, (October, 1947), p. 56.

Loving, D., "Solarize Negatives." *American Photography*, (April, 1943), p. 16.

Tennen, K. and S. Levine, "A Recipe For Solarization." *Petersen's Photographic Magazine*, (June, 1973), p. 26.

Darkroom and print finishing techniques not restricted to solarizing:

Croy, O. R., *The Complete Art Of Printing And Enlarging*, New York, Focal Press, 1959.

———*Design By Photography*, New York, Focal Press, 1963.

———*Creative Photography*, New York, Focal Press, 1965. These three books by Dr. Croy constitute an excellent course in darkroom manipulations, probably the best available.

Fraprie, F. R. and W. E. Woodbury, *Photographic Amusements*, Garden City, Amphoto, 1931.

Gibbs, C. W., "Paper Negatives." *American Photography*, (October, 1934).

Mortensen, W., *Print Finishing*, California, Camera Craft, 1938. Introduces the technique of abrasion tone control.

———*The Paper Negative*, California, Jay Curtis, Co., 1954. "Pictures Made In The Darkroom." *Popular Photography*, (October, 1949), p. 76.

Ward, N., *Picture Making With Paper Negatives*, Garden City, Amphoto, 1931.

Color manipulations including solarization:

Fontaine, A., "Tips From Pros." *Popular Photography*, (October, 1951).

Gareis and Scheerer, *Creative Color Photography*, Hastings-on-Hudson, Fountain Press, 1970. Includes the most extensive discussion of solarization of color materials.

Heyduk, M., "Creative Color Pseudo-Solarization." *Camera 35*, (October/November, 1967), p. 52.